ENGINEERING DRAWING

PROBLEM SERIES 2
TENTH EDITION

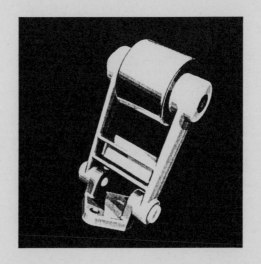

Frederick E. Giesecke / Alva Mitchell / Henry Cecil Spencer
Ivan Leroy Hill / Robert Olin Loving
John Thomas Dygdon / James E. Novak

Prentice Hall
Upper Saddle River, New Jersey 07458

Acquisitions Editor: *Eric Svendsen*
Editorial Assistant: *Andrea Au*
Production Editor: *Dawn Blayer*
Production Coordinator: *Donna Sullivan*
Special Projects Manager: *Barbara A. Murray*
Cover Designer: *Paul Gourhan*

 © 1997 by **PRENTICE-HALL, INC.**
Simon & Schuster/A Viacom Company
Upper Saddle River, NJ 07458

Printed in the United States of America

10 9 8 7 6 5 4 3 2

ISBN 0-13-658881-6

Prentice-Hall International (UK) Limited, *London*
Prentice-Hall of Australia Pty. Limited, *Sydney*
Prentice-Hall Canada, Inc., *Toronto*
Prentice-Hall Hispanoamericana, S.A., *Mexico*
Prentice-Hall of India Private Limited, *New Delhi*
Prentice-Hall of Japan, Inc., *Tokyo*
Simon & Schuster Asia Pte. Ltd., *Singapore*
Editora Prentice-Hall do Brasil, Ltda., *Rio de Janeiro*

Preface

This workbook may be used with any good reference text. Since the time available for the teaching of technical drawing is limited, the objective has been to create a collection of problem sheets that eliminate repetitious drawing and still give adequate coverage of the fundamentals. It is expected that in many cases the instructor will supplement these problem sheets with assignments of problems from the text to be drawn on blank paper.

Most of the problems in this workbook are taken directly from industry, and are carefully selected from thousands of original drawings in drafting department files. Although a given print from industry may in itself be too simple or too complicated for classroom use, it is useful as a basis for the classroom problem. There is no problem in drawing that does not have its counterpart on prints used in industry. By means of many notes, supplementary drawings, and pictorials, every effort has been made to connect the problems to the practical situations from which they were taken.

An outstanding feature of this revision is that fractional-inch dimensions have largely been replaced by the decimal-inch and metric dimensions now used extensively in industry. A decimal and millimeter equivalents table and appropriate full-size and half-size scales are provided inside the front and back covers for the student's convenience.

Technical sketching is recognized as an important skill in engineering work, and an entire unit on multiview and isometric sketching is included early in the book. Grids are provided on the first sheets to help the student gain confidence. The instructor may assign many of the problems to be drawn freehand for additional emphasis of this technique.

All sheets are 8.5" x 11.0" in conformity with American National Drafting Standards for engineering drawing practices, a size that facilitates handling and filing by the student and the instructor. Several problems are printed on vellum to provide experience in the manner of commercial practice. It is expected that in most cases the instructor will supplement these problem sheets with assignments of problems from the text to be drawn on blank paper, vellum or film.

All of the problems are based upon actual industrial designs, and their presentations are in accord with the latest ANSI Y14 American National Standard Drafting Manual and other relevant ANSI standards.

In response to the increased usage of computer technology for drafting and design, a number of problem sheets in computer-aided drafting (CAD) have been included. The problem sheets on detail drawings are presented to provide practice in making regular working drawings of the type used in industry. These are suggested for solution either by a computer-aided drafting system or by traditional drafting methods.

A special feature of the workbook is the Instructions, in which a detailed explanation of the requirements of each problem is given, together with specific references in the text to cover each point. These instructions are expected to anticipate many of the questions that will arise in the student's mind in connection with each problem, thereby freeing the instructor from much detailed individual instruction and providing more time for aid to students who have basic difficulties.

Most of the problems have been used extensively in engineering colleges throughout the country and, therefore, are well tested under classroom conditions. The cooperation of many schools and companies in supplying the original prints for these problems is deeply appreciated. The valuable contributions of Prof. H. E. Grant, originally a co-author of this book, are gratefully acknowledged.

Comments and criticisms from the users of this workbook will be most welcome.

Ivan Leroy Hill
Clearwater, FL

John Thomas Dygdon
*Illinois Institute
of Technology
Chicago, IL*

James E. Novak
*Illinois Institute
of Technology
Chicago, IL*

Contents

Vellums

Instructions

Alternative dimensions, often not the *exact* equivalents, are given in millimeters and inches. Although it is understood that 25.4 mm = 1.00", it is more practical to use approximate equivalents such as 25 mm for 1.00", 12.5 or 12mm for .50", 6 mm for .25", 3 mm for .12", and so on. Exact equivalents should be used when accurate fit or critical strength is involved.

In general, the following leads are suitable for instrumental drawing: a 4H for construction lines and guide lines for lettering; a 2H for center lines, section lines, dimension lines, and extension lines; and an HB or F for general linework and lettering. Instructions for specific drawings may suggest the use of other grades of lead as required. All construction lines on problems should be made *lightly* and *should not be erased.*

Drawing 1-1 Instrumental Drawing: Alphabet of Lines.
Spaces 1 and 2. Draw indicated lines full length in the given spaces.
Space 3. Section lines at 30° with horizontal upward to the right and 12.5 mm (.50") apart are required. First find center of space by drawing diagonals in *very light* construction lines. Through center draw construction line 60° with horizontal upward to the left and, beginning at the center, set off 12.5 mm (.50") intervals. Draw required section lines through these points to fill space.
Space 4. Using similar construction methods to those outlined for *Space 3*, draw visible lines at 75° with horizontal upward to the left at 12.5 mm (.50") intervals.
Space 5. Draw center lines parallel to given line and at 12.5 mm (.50") intervals to fill the space.
Space 6. Draw alternate visible and hidden lines 12.5 mm (.50") apart and perpendicular to the given line. Arrange so that one visible line passes through center of space.

Drawing 1-2 Instrumental Drawing: Scales and Layout.
Space 1. Use architects', engineers', or metric scales as necessary. Measure lines A, C, and F at the scales shown, and indicate the scale lengths (L) at the right. At B, D, E, G, and H draw lines of specified lengths at scales shown. Terminate the lines in the same manner as for given lines. At J through L, determine the scales and lengths of lines and record the scales and lengths in the spaces provided.

Line J is over 500' and under 600' in length.
Line K is between 530 m and 550 m in length.
Line L is one twenty-fourth size.
Space 2. Draw the two views of the Anchor Bracket full size, locating the views by the starting corners indicated.

Drawing 1-3 Alphabet of Lines. In order to learn the correct technique of drawing all the common lines in technical drawing, and the proper order of penciling a drawing, you are to make an exact copy of the drawing shown, drawing the circular view at center A, and omitting all instructional notes shown in *inclined* lettering. The only lettering that you are to copy is the 200 mm dimension and the finish mark. On your drawing, carefully use all measurements indicated, such as size of arrowheads, height of dimension numerals, etc. Do not give dimensions for these on your drawing.
First. Draw center lines using a sharp 2H lead. Block in the two views with extremely light construction lines using a sharp 4H lead. For the arcs, use your bow pencil with a sharp 4H lead. Adjust needle point with shoulder end out. Use your bow dividers to take arc radii from the given drawing, and then draw the arcs *lightly* from center A below. Next, draw straight horizontal and vertical construction lines in the circular view and then in the right-hand view. Note that horizontal lines can be projected across from view to view and that vertical lines can be projected down from the drawing above.
Second. Heavy in the views and finish the drawing, using a sharp F lead in your bow pencil for the

1

arcs. All final lines should be *clean* and **dark.** Use both hands on the bow pencil for all circular hidden dashes in order to control the lengths of the dashes. Use the large compass only for arcs of more than 25 mm (1.00") radius.

Third. Draw extension lines and dimension line **dark** but *very sharp,* using a 2H lead. Draw horizontal and vertical guide lines *lightly* for dimension numerals with a 4H lead. Do not erase construction lines. They should be so light that they are barely visible.

Drawing 2-1 Vertical Lettering: Capitals and Numerals. Using an HB lead, letter the indicated characters in the spaces provided. These large letters and numerals may be sketched lightly first, and then corrected where necessary before being made heavy with the strokes shown. Omit the numbers and arrows from your letters. All lettering must be *clean-cut* and **black.** In the title strip, under DRAWN BY, draw light guide lines from the starting marks shown, plus random vertical guide lines, and letter your name with the last name first. Under FILE NO. letter your identification symbol as assigned by your instructor.

Drawing 2-2 Inclined Lettering: Capitals and Numerals. Using an HB lead, letter the indicated characters in the spaces provided. These large letters and numerals may be sketched lightly first, and then corrected as necessary before being made heavy with the strokes shown. Omit the numbers and arrows from your letters. All lettering must be *clean-cut* and **black.** In the title strip, under DRAWN BY, draw light guide lines from the starting marks shown, and letter your name with the last name first. Under FILE NO. letter your identification symbol as assigned by your instructor.

Drawing 2-3 Vertical Lettering: Capitals and Numerals. First draw light vertical guide lines at random from bottom to top of the sheet. Do not draw separate vertical guide lines for each line of lettering. Reproduce the lettering as exactly as you can, using an HB lead for the larger letters and a sharp F lead for the smaller letters. Note that the last line of lettering is to be lettered twice. All letters must be *clean-cut* and **black.**

Drawing 2-4 Inclined Lettering: Capitals and Numerals. First, draw light inclined guide lines at random from bottom to top of the sheet. Do not draw separate inclined guide lines for each line of lettering. Reproduce the lettering as exactly as you can, using an HB lead for the larger letters, and a sharp F lead for the smaller letters. Note that the last line of lettering is to be lettered twice. All lettering must be *clean-cut* and **black.**

Drawing 2-5 Vertical Letter: Shop Notes and Dimensions. On the left side of the sheet are shown a number of lettering applications. On the right, re-produce the lettering, arrowheads, and finish marks, using a sharp F lead. Except for the title TOOL HOLDER at the bottom, all lettering on the sheet is 3 mm or .12" high. Draw all guide lines with the aid of a lettering triangle or the Ames Lettering guide, using a 4H or 6H lead. Letter, in vertical capitals, the title TOOL HOLDER, etc., on center. Use height and spacing as specified. Underline TOOL HOLDER. All lettering must be *clean-cut* and **black.**

Drawing 2-6 Inclined Lettering: Shop Notes and Dimensions. On the left side of the sheet are shown a number of lettering applications. On the right, repro-duce the lettering, arrowheads, and finish marks, using a sharp F lead. Except for the title TOOL HOLDER at the bottom, all lettering on the sheet is 3 mm or .12" high. Draw all guide lines with the aid of a lettering triangle or the Ames Lettering Guide, using a 4H or 6H lead. Letter the title TOOL HOLDER, etc., on center. All lettering must be *clean-cut* and **black.**

Drawing 3-1 Geometric Construction: Drafting Geometry. Show light construction on all problems. Add center lines where necessary.

Space 1. Locate and draw hole as indicated. Add center lines.

Space 2. Complete the view of the Special Washer.

Space 3. Complete the view of the Rack. Start the first tooth at A as indicated.

Space 4. Locate centers for holes as specified. Draw holes and add center lines.

Space 5. Complete the view of the End Guide by adding the 90° tip as indicated in the given specifications.

Space 6. Complete the view of the Bracket.

Drawing 3-2 Geometric Construction: Tangencies. Draw all construction lines *lightly* with a sharp 4H lead, and do not erase. Draw all required lines with a sharp F lead, **dark** and *clean* to match the given lines. *Show all points of tangency* by means of light construction lines.

Drawing 3-3 Geometric Constructions: Polygons, Elipses, and Parabolas. Show light construction on all problems.

Space 1. Complete the end view of the 12-point (double hexagon) socket as indicated.

Space 2. Draw the view of the octagonal face only of the engineers' cross-peen hammer head, which is cut from 28 mm square stock.

Space 3. Determine diameter of milling cutter to cut true arc through points A, B, and C. Give diameter to nearest 0.5 mm.

Space 4. Using the concentric-circle method, draw profile of 56 mm x 82 mm elliptical cam, starting at point A. Use a minimum of 16 points to establish the ellipse. Draw ellipse with the aid of the irregular curve.

Space 5. Draw approximate ellipse as indicated.

Space 6. Draw parabola as indicated, and find focus.

2

Drawing 4-1 Normal Surfaces. Follow instructions on the sheet. Omit dimensions unless assigned. Study the illustration at the right of the isometric and be able to explain the function of the Necking Tool Post. If your lines are not **black**, a clear print cannot be made from your drawing.

Drawing 4-2 Inclined Surfaces. Follow instructions on the sheet and draw the required views. Use a 4H lead for construction lines (extremely *light* — do not erase) and a sharp F lead for visible lines and hidden lines. Add numbers to the front view as indicated, making them small to match those in the side view. Note that the numbered inclined surface has the same number of edges in the side and front views. The dimension 10 X 7 indicates that the groove is 10 mm wide and 7 mm deep. Use dividers for projecting from the side view to the top view.

Drawing 4-3 Combination Edges and Surfaces. Draw with instruments to the scale indicated in the top and side views, showing all hidden lines and omitting dimensions. Make fillets and rounds very carefully to match those shown in the front view. Use the bow pencil, sharpened and carefully adjusted for drawing the larger fillets and rounds. Draw the smaller ones carefully freehand to match those given. Use a sharp F lead, and make all lines *clean* and **dark** so that a clear print can be made from your drawing. Add necessary finish marks.

Drawing 5-1 Technical Sketching: Multiview. Sketch the views as indicated. Make all final lines *clean-cut* and **black**. In the space provided, letter the names of the necessary views.

Drawing 5-2 Techinal Sketching: Multiview and Isometric. Using an HB lead, sketch the views or isometrics as indicated. Make lines *clean-cut* and **black**. Omit hidden lines from the isometrics. Take great care to sketch isometric ellipses correctly.

Drawing 6-1 Multiview Projection: Missing Lines. Each problem is incomplete because lines are missing from one or more views. Add all missing lines freehand, including center lines.

Drawing 6-2 Multiview Projection: Missing Views. In each problem two complete views are given, and a third view is missing. Add the third view in each case freehand.

Drawing 6-3 Multiview Projection: Missing Views and Lines. For probs. 1-3, show constructions for the points of tangency. Draw all necessary center lines.

Drawing 6-4 Multiview Projection: Missing Views. In each problem two complete views are given. Add the third view in each case, using instruments.

Drawing 6-5 Multiview Projection: Missing Views. *Space 1.* Draw freehand the top view as indicated.

Space 2. Draw with instruments the front view as indicated. Plot a sufficient number of points to define the curve accurately. Sketch the curve lightly through the points with a 4H lead; then draw the curve with the F lead and with the aid of the irregular curve.

Drawing 7-1 Computer-Aided Drafting: Terms and Descriptions. Some terms related to computer graphics are given in the table. A list of descriptions for these terms is given on the right. Find the matching description for each term and enter its letter identifier in the table.

Drawing 7-2 Computer-Aided Drafting: Two-Dimensional Coordinate Plot.
Space 1. Digitize the single view drawing by defining the X and Y coordinates of the indicated points and fill in the given table. Point A is the origin with values of X and Y equal to zero. Consider each division of the grid as 1 unit. Keep in mind that any X values to the left of the origin and any Y values below the origin are negative.
Space 2. From the X and Y coordinate data given in the table, plot all points on the grid and complete the drawing. Point A is the origin. Consider each division of the grid as 1 unit.

Drawing 7-3 Computer-Aided Drafting: Three-Dimensional Coordinate Plot. In drawing an image, the actions of the pen are Move and Draw.

Move: The pen moves from its present position to new X, Y, and Z coordinates specified. A line is not drawn. Numeral 0 is used to indicate Move action.

Draw: A line is drawn from the present pen position to new X, Y, and Z coordinates specified. Numeral 1 is used to indicate Draw action.

Space 1. Determine X, Y, and Z coordinates for all the points of the object. Complete the table for drawing the object, starting with point A. Coordinates X, Y, and Z are positioned as indicated by the arrows, with point A as origin. Try to use a minimum number of Move actions.
Space 2. According to the data shown in the table, draw the object on the grids provided. Coordinates X, Y, and Z are positioned as indicated by the arrows, with point A as origin.

Drawing 7-4 Computer-Aided Drafting: Menu Usage. The drawing shows the front view of a Bracket that is to be generated on a graphics terminal. The numbers 1 to 21 refer to graphic entities that make up the drawing. Available menu commands for generating entities are given on the right. Complete the table by determining the menu commands to generate the entities. Enter the letter identifiers (A, B, etc.) of menu selections in the table.

Drawing 7-5 Computer-Aided Drafting: Coordinate Systems. Using the given descriptions for VIEW COORDINATES and WORLD COORDINATES, complete the tables for the front and right-side views of the object. Point number 1 is considered as the origin. Each grid division is equal to 1 unit.

Drawing 8-1 Sectional Views: Full and Half. Sketch the sections as indicated, using an HB or F lead for visible lines and a sharp F lead for section lines and center lines. Make section lines thin to contrast well with heavy visible lines. All lines should be *clean-cut* and black. No additional cutting planes are required.

Drawing 8-2 Sectional Views: Full and Revolved. Draw the indicated sectional views, using insruments. Use an F lead for visible lines and a *very sharp* 2H lead for section lines. Add center lines. Omit finish marks unless assigned. In Space 1, include a revolved partial section in the front view to show the shape of the triangular rib. See Fig. 9.16 (m). In Space 2, fillets are 3 mm (.12") R.

Drawing 8-3 Sectional Views: Full and Removed. Draw sections as indicated, including visible lines behind the cutting planes. Add center lines. If assigned, add finish marks.

Drawing 8-4 Sectional Views: Full and Assembly. Draw the indicated sectional views, using instruments. In Prob. 2 is shown a portion of an assembly full section with a round shaft extending through a cast-iron cover and a steel plate, which are held together by bolts. Section-line the sectioned areas, using symbolic section lining for each material.

Drawing 8-5 Sectional View: Aligned. Draw Section A—A as indicated.

Drawing 8-6 Full-Section Views. Draw full sections, using a sharp F lead for visible lines, and a sharp 2H lead for section lines and center lines. Omit all hidden lines. Make fillets and rounds carefully with the bow pencil to match those in the given views. Space section lines about 4 mm (.16") apart. Add finish marks to sectioned views. In a sectioned area, omit the section lines where the finish mark is placed. Drilled holes, counterbored holes, reamed holes, etc., are understood to be finished; hence, finish marks are omitted.

Drawing 8-7 Broken-Out and Removed Sections. Draw required sections. Use a fairly sharp F lead for visible lines and a sharp 2H lead for section lines and center lines. Draw break line in Prob. 1, then draw the broken-out section. In the removed sections, show all visible lines behind the cutting plane in each case. Carefully draw the small fillets and rounds freehand. Add finish marks in both problems.

Drawing 8-8 Aligned Sections. Use a fairly sharp F lead for visible lines and a sharp 2H lead for section lines and center lines. Add finish marks in the sectional views.

Drawing 9-1 Auxiliary Views: Primary. Sketch the auxiliary views as indicated. Using an HB or F lead, make visible lines and hidden lines black so that the views will stand out clearly from the grids. Use a sharp F lead for center lines. Letter folding lines. In Probs. 3 and 4, include all hidden lines.

Drawing 9-2 Auxiliary Views: Primary. Using insruments, draw the indicated views. Use folding lines in Prob. 1 and reference-plane lines in Probs. 2–4. Prob. 3, dimension the angles between surfaces A and B, and A and C.

Drawing 9-3 Auxiliary Views: Primary. Using instruments, add any missing lines in the regular views or auxiliary views.

Drawing 9-4 Auxiliary Views: Secondary. Using instruments, draw the indicated auxiliary views.
Space 1. Surface A is a normal surface, and therefore appears true size in the given view. Use reference-plane lines. Dimension the 135° angle.
Space 2. The height of the object is shown in the reduced-scale drawing. Draw the primary auxiliary view 22 mm from the given top view. Use folding lines. Dimension the angle between surfaces A and B in degrees.

Drawing 9-5 Auxiliary Section. Draw required auxiliary section, using a fairly sharp F lead for visible lines and a sharp 2H lead for section lines and center lines. Note that the edges of the cored hole are rounded. Observe that in the auxiliary view the object will be in an inverted position with the bottom of the object *up*. Show finish marks. The smaller rounded and filleted edges should appear as conventional edges in the auxiliary view.

Drawing 10-1 Revolutions: Primary. Using instruments, draw the indicated constructions. In Probs. 1 and 2, use alternate position lines to show the revolved surfaces.

Drawing 11-1 Dimensioning: Freehand. Use the complete decimal dimensioning system with metric values. If assigned, use decimal-inch equivalents. Add dimensions freehand, spacing dimension lines approximately 10 mm from the views and 10 mm apart. Include necessary finish marks in Prob. 2. Note that in Prob. 2 the drawing is half the size of the actual part. The two small holes are drilled, Appendix 16, and the large hole is bored. In the bored-hole note, specify the diameter to two decimal places.

Drawing 11-2 Dimensioning: Mechanical. Measure the views (from centers of lines), and dimension completely, using the complete decimal dimensioning system with metric values to the nearest 0.5 mm (.02"). If

4

assigned, use decimal-inch dimensions. Notice the scales in each problem. Space dimensions uniformly 10 mm (.40") from the object and 10 mm (.40") apart. Use the unidirectional system. Draw light guide lines for all dimension figures and notes. Use a sharp 2H lead for dimension lines, extension lines, and center lines; a sharp F lead for arrowheads and lettering. Give radii dimensions for the larger arcs, and give notes covering the small fillets and rounds. Use V-type finish marks in Spaces 1 and 2. In Space 2 give the angle in degrees. The large hole is reamed for an RC 6 fit. Give the ream note in millimeters rounded off to two decimal places. *Caution:* Put dimensions in place before lettering notes; then choose the best open spots for the notes and letter them without crowding.

Drawing 11-3 Dimensioning: Mating Parts. The T-Slot Clamp is a holding device used in the machine shop to hold work pieces in position. Dimension the parts completely, spacing dimensions lines 10 mm from the views and 10 mm apart. Use the complete decimal dimensioning system with metric values. If assigned, use decimal-inch equivalents. Note that finish marks are not required on parts made of CRS (cold rolled steel).

Part No. 1. Frame. The hole in the base is drilled 1.5 mm (0.6") larger than the M10 X 1.5 T-Slot Bolt. The bottom of the base is machined flat. For the tapped hole use metric coarse threads or if assigned, use Unified Coarse threads.

Drawing 11-4 Dimensioning: Mating Parts (continued).
Part No. 2. Clamp Screw. The part is cylindrical except for the spherical ball on the end. The threads correspond to those in the Frame. The length of the threaded portion, including the chamfer and relief, is 66 mm. The drilled hole is 0.6 mm larger than the nominal size of the Handle.

Part No. 3. Pad. The drilled hole is 0.5 mm larger than the ball of the Clamp Screw. The Pad is attached to the ball of the Clamp Screw by crimping the slit edges.

Part No. 4. Handle. The Handle is 90 mm long. The diameter is 6.375–6.385 mm.

Part No. 5 Handle Cap. The Cap is cylindrical. The hole is drilled and reamed to 6.350-6.365 mm diameter so as to fit tightly on the Handle.

Drawing 11-5 Limit Dimensioning. Dimension the details completely. Use the unidirectional metric or decimal-inch system as assigned. Use two-place inch decimals or one-place millimeter decimals for all except the limit dimensions specified. Use complete guide lines, making whole numbers and notes 3 mm (.12") high.

The required American National Standard fits are indicated on the small assembly. For fit "A" give tolerances of 0.05 mm (.002") and an allowance of 0.08 mm (.003"), with the allowance subtracted from the hub and the bushing. The handwheel, shaft, and pinion should revolve freely where the nuts are securely tightened.

For Woodruff keyseats, give one note: NO. 505 WOODRUFF KEYSEATS. For threads, give one note leading to either end of front view, but add: BOTH ENDS. Use equivalent metric threads or Unified Fine Series Threads. For chamfers give one note but add: BOTH ENDS. For fillets, give one radius.

Drawing 12-1 Threads and Fasteners: Nomenclature and Identification. Draw light guide lines from the marks indicated, and letter the answers in the spaces provided. Standard abbreviations may be used to avoid crowding.

Drawing 12-2 Threads and Fasteners: Schematic Symbols. Draw specified threads and fastener details, using the schematic thread symbols unless otherwise assigned. Complete the section lining and leaders where required. Chamfer ends of threads 45° x thread depth in Probs. 1–3.

Drawing 12-3 Threads: Detailed. Using detailed representation, show the specified threads for the given sectioned assembly of the cylindrical Coupler. The internal thread of the Core is a through thread and the stud of the Piston Rod is engaged to the depth indicated. Complete the section lining and the thread-note leaders.

Drawing 12-4 Acme and Square Threads: Detailed.
Adjusting Screw. Draw the specified Acme threads to complete the view. Complete the leaders and add arrowheads touching the threads. Construct the threads so as to be symmetrical about the central neck of the screw.

Leveling Jack. Draw the specified square threads to complete the assembly view. Note that the scale of the drawing is double size. Add necessary seciton lining and complete the thread-note leader.

Drawing 12-5 Square Threads. Draw square threads as indicated. Complete all section lining, using a sharp 2H lead, but do not section the Jack Screw. Section the Jack Screw Nut up to the break line near the top.

Drawing 12-6 Fasteners.
Prob. 1. Select lock washer next size larger than shank of screw. Complete the assembly in section, using CI section lining for all parts.

Prob. 2. Complete the assembly in section, using CI section lining for all parts. Do not section screws.

Drawing 13-1 Isometric Drawing: Freehand.
Spaces 1 to 6. Sketch isometrics of given objects. Note the given starting point A.

Drawing 13-2 Isometric Drawing: Mechanical. Omit hidden lines in all problems. All "box construction" and other construction lines should be made lightly with a sharp 4H lead, and should not be erased. Darken all visible lines with a sharp F lead.

Spaces 1–3. Draw isometric drawings of the objects shown, locating the corners at A and using the dividers to transfer distances from the views to the isometrics.

Space 4. Draw isometric drawing, locating the corner A at the point A. Use the scale to set off dimensions. Do not transfer distances with dividers, as the given drawing is not to scale. Show construction for the 30° angle.

Space 5. Complete the isometric drawing, using the information supplied in the reduced-scale drawing.

Space 6. Complete the isometric drawing, transferring measurements directly from the given views to the isometric with dividers. Draw the final curves with the aid of the irregular curve.

Drawing 13-3 Isometric Drawing: Mechanical. Using instruments, make isometric drawings of two assigned problems. Use the indicated starting corners. Show all construction.

Drawing 13-4 Isometric-Irregular Combinations. Draw isometric of either assigned problem, taking dimensions from the given views with dividers, and doubling them on the isometric. Omit hidden lines unless necessary for clearness.

Drawing 14-1 Oblique Projection: Freehand. Using the method shown in Fig. 6.28, make oblique sketches of objects shown, using the starting corners indicated. Omit hidden lines and center lines. Make visible lines **black** so sketches will stand out clearly from the grids.

Drawing 14-2 Oblique Projection: Cabinet and Cavalier. Draw assigned problem in Space 1 in cavalier projection and in Space 2 in cabinet projection. Omit hidden lines unless necessary for clearness. Show all constructions clearly, and do not erase construction lines. In Space 2, Prob. 1 (a), the ellipse construction must be drawn by means of offsets.

Drawings 15-1 to 18-2

Accuracy. Graphic solutions to space problems require accurate measurements and clean, sharp line work. Properly sharpened F and 2H leads will be found suitable for most line work, while a sharp 4H lead is preferred for construction and guide lines for lettering. The F lead is normally used for lettering. Make all measurements from center to center of lines and when no scale is specified, measure to the nearest 0.2 mm.

Unless otherwise specified, the basic unit for the given scales is the millimeter. Hence, a scale of 1/2000 is 1 mm to 2000 mm or 2 m. For measurements equal to or greater than 1000 mm, indicate these measurements in meters.

Notation. A certain amount of lettering is necessary on all graphical solutions. A minimum amount of notation should include the following:

1. Label at least one point in each view, or label points that are mentioned in the instructions.
2. Label all folding lines employed.
3. Show the symbols for EV (edge view), TL (true length), TS (true size), and LI (line of intersection) when they are a part of the solution.
4. Show given and required information such as angles, distances, bearings, and other numerical items *on the views* where measured or set off.

Drawing 15-1 Points and Lines: Visibility. Use standard alphabet of lines to complete the solutions. Change the dotted lines to standard lines in Spaces 3 and 4.

Drawing 15-2 Points and Lines: Points on Lines.

Space 1. Point 2 is 33 mm to the right of point 1, 25 mm below point 1, and 20 mm in front of point 1. Show these dimensions on the drawing.

Space 2. Line 1–2 is 38 mm long (2 is behind 1). Line 1–3 is a 40 mm long frontal line. Show these dimensions on the drawing. Line 2–3 is a profile line. Dimension the true length of line 2–3.

Space 3. Line 1–2 is 33 mm long. Show this on the drawing. The front view of line 2–3 is true length as indicated.

Space 4. Point 5 is on line 1–2 and is 15 mm above point 1. Point 6 is on line 3–4. Line 5–6 is a horizontal line. Dimension the length of line 5–6.

Space 5. Note that point 4 is to be moved vertically *in space.*

Space 6. Triangle 1–2–3 is the base of a pyramid. Vertix V is 5 mm behind point 1, 10 mm to the left of point 2, and 28 mm above point 3. Show these dimensions on the drawing.

Drawing 15-3 True Length of Line: Angles by Auxiliary Views. Use auxiliary views. Dimension the required angles in degrees and also dimension the true lengths in millimeters for Spaces 1–3. Show all given data on the drawing for Space 4.

Drawing 15-4 True Length of Line: Bearing and Grade. Use auxiliary views. For all problems indicate the percent grade on the drawing. For Spaces 1–3, also show the numerical values of the bearing and true length on the drawing.

Drawing 15-5 True Length of Line: Revolution. Use American National Standard phantom lines for lines in alternate or revolved-position views.

Space 1. Dimension the true length, slope, and bearing on the drawing.

Space 2. Dimension ∠F, ∠P, and the true length on the drawing.

Space 3. Show the 30° angle on the drawing.

Space 4. Show all given data on the drawing.

Drawing 15-6 True Length of Line: Point View.

Space 1. Include the minimum notation as recommended in the general instructions for Drawings 19-1 to 22-2.

6

Space 2. Dimension the clearance on the drawing.

Space 3. Dimension the true distances for comparison.

Drawing 15-7 Planes: Points and Lines in Planes.

Space 1. Indicate your answer in the space provided.

Space 2. The *top* view of point 4 and the *front* view of point 5 are shown.

Space 4. Show given data on the drawing.

Drawing 15-8 Planes: True Size.

Space 1. Indicate values for calculation of the area on the drawing and show your calculations. Record the area to two significant figures only, in the space provided.

Space 2. Show the connecting feeder line in the given views.

Drawing 15-9 Lines and Planes: Piercing Points—Edge-View Method.

Use the edge view method on this sheet and show proper visibility in all views. Show EV where appropriate and encircle the piercing points in all views.

Space 3. Do not show a hidden line for that segment of line 1–2 which is within the pyramid.

Drawing 15-10 Lines and Planes: Piercing Points—Cutting-Plane Method.

Use the cutting-plane method on this sheet and show proper visibility in all views. Show EV on the lines as edge-view cutting planes and encircle the views of the piercing points.

Space 3. Do not show a hidden line for that segment of line 1–2 which is completely within the pyramid. However, include any extension lines necessary to clarify the method of solution.

Space 4. Note that the given views are a front and a primary auxiliary view.

Drawing 15-11 Planes: Intersections.

Show the symbol EV wherever appropriate and label the line of intersection with the symbol LI.

Spaces 1 and 2. Show complete visibility.

Spaces 3 and 4. Visibility is of no concern in these problems, but special care is needed to insure accuracy of your solutions.

Drawing 15-12 Planes: Dihedral Angles.

Space 1. Dimension the angles on the drawing. Show complete visibility.

Space 2. Special care is needed to assure accuracy in your solution of this problem. Show the angle on your drawing. Show complete visibility.

Drawing 15-13 Lines and Planes: Angle Between Line and Plane.

Space 1. Dimension the angle on the drawing.

Space 2. This is an "open ended" problem (as are most engineering problems) in that there are many possible answers. Show use of given data on the drawing.

Drawing 16-1 Parallelism: Lines and Planes.

Space 1. Demonstrate your answer graphically and indicate it by a check mark(s) (✔) in the space provided.

Space 2. You may check vertical alignment with a T-square and a triangle, but do not perform any actual constructions. Indicate your answer(a) by a check mark (✔) in the spaces provided.

Spaces 3 and 4. Use only the given views.

Drawing 16-2 Parallelism: Lines and Planes.

Space 1. Show bearing on the drawing.

Drawing 16-3 Perpendicularity: Lines.

These problems are to be solved without the construction of auxiliary views. Use the symbol to show lines drawn perpendicular.

Drawing 16-4 Perpendicularity: Lines and Planes.

Space 1. Do not construct additional views.

Space 2. Use an auxiliary view, or only the given views, or both, as assigned. Show the length of the altitude on the drawing.

Space 3. Use auxiliary views. Show *all* views in completed form with proper visibility.

Drawing 16-5 Skew Lines: Common Perpendicular.

These problems are designed to be solved by the point-view method. Use of the plane method is not recommended because of space limitations.

Space 1. Show answers on views where measured. Be sure to check accuracy with divider distances.

Space 2. Indicate the actual clearance distances on the drawing where measured. Show the front and side views of lines representing the two distances. Indicate with a check mark (✔) in the spaces provided, the answer to the matter of inspection.

Drawing 16-6 Skew Lines: Lines Specified Angles.

Spaces 1 and 2. Show given data on the drawing where measured.

Drawing 17-1 Intersections: Planes and Polyhedra.

Space 1. The edge-view method is suggested. Be sure to check accuracy with divider distances. Show visibility.

Space 2. Use the cutting-plane method, including the EV symbols. Show complete visibility except that hidden line segments of the plane's boundaries within the prism should be omitted.

Drawing 17-2 Intersections: Prisms and Pyramids.

Omit hidden line segments of lateral edges of either solid which are within the other solid.

If assigned: On a separate sheet develop the surface of one or more of the solids, including the intersections.

Drawing 17-3 Intersections: Circular Forms.

Determine the figures of intersection and show visibility.

Drawing 17-4 Parallel-Line Development: Prism and Cylinder.

Show full developments inside up. Omit the bases.

Space 2. Calculate the length of the development for the cylinder. Start development with the shortest element.

Drawing 17-5 Radial-Line Development: Pyramid and Cone. Show developments inside up. Start with shortest seam. Omit bases.

Drawing 17-6 Triangulation: Transition Piece. Show half development inside up. Omit the cylindrical and prismatic connectors at the top and bottom.

Drawing 18-1 Tangencies: Line and Plane. If assigned, where alternative solutions are possible, show the second solution with a phantom line.

Drawing 18-2 Tangencies: Specified Angles. If assigned, show alternative solutions with phantom lines.

Drawing 19-1 Graphs: Pie Chart and Bar Graph.
Space 1. According to the U.S. census (1970), Chicagoans age 25 and over (1,943,464) reported the following educational achievement:

Never attended school	2%
Dropped out before grade 8	18%
Dropped out before grade 9	16%
Dropped out in high school	21%
Graduated from high school	26%
Dropped out in college	9%
Graduated from college	8%

Divide the pie chart to illustrate the above data. Show and label the appropriate percentages for each category. Use 3 mm (.12") engineering lettering throughout. Balance the largest sector symmetrically about a vertical center line in the lower area of the circle. Title the chart: EDUCATIONAL EXPERIENCE OF CHICAGO RESIDENTS AGE 25 AND OVER—1970. Underline the title only. Indicate outside the area of the pie chart the total number that reported.
Space 2. Construct a column or bar chart on the rectangular grid to show the auto accidents experienced by drivers of various age groups. The Motor Vehicle Mfrs. Assn. of U.S., Inc., reported the following data (1972):

Age Groups	Licensed Drivers	Drivers Involved in Fatal Accidents
Under 25	21.6%	33.8%
25–44	39.3%	37.7%
45–64	30.6%	21.1%
65 and over	8.5%	7.4%

Locate the axes at the given starting corner. Place the PERCENT scale on the *y*-axis and the AGE GROUP scale along the *x*-axis. Two bars are required for each age group. Cross-hatch at 45° the bars representing the licensed drivers. Indicate the percent values above all bars. Add border lines for the chart area and strengthen appropriate horizontal grid lines, but do not cross bar areas. Identify shading significance with key inserts. Title the chart: INVOLVEMENT IN FATAL ACCIDENTS, U.S. DRIVERS—1972.

Drawing 19-2 Graphs: Pie Chart and Bar Graph.
Space 1. Worldwide production of cars, trucks, and buses for 1979 totaled 41,515,000 units. The percentage distribution was as follows:

U.S.	28%
Japan	23%
W. Germany	10%
France	9%
U.S.S.R.	5%
Italy	4%
Canada	4%
G. Britain	3.6%
All others	13.4%

Divide the pie chart to illustrate the given data. Label and show the value for each sector. Use 3 mm or 0.12" engineering lettering. Place the sum of the two largest sectors about a vertical center line in the lower portion of the circle.
Space 2. Construct a bar chart for the following data on nuclear power generation.

Nuclear Power Generation
(Millions of kilowatt-hours)

1973	85,000
1974	115,000
1975	170,000
1976	190,000
1977	255,000
1978	280,000
1979	260,000
1980	275,000

Use 12 mm wide vertical bars beginning at the given horizontal base line. Allow 6 mm spaces at the beginning and end and between the bars. Shade the bars. Title the chart in 3 mm or 0.12" letters. Draw horizontal grid lines for each 50,000 value indicated, but do not draw them across the bars.

Drawing 20-1 Detail Drawings. Draw or sketch the necessary views of the object assigned. Select appropriate scale and sheet size. Dimension completely using metric or decimal-inch dimensions as assigned.
Alternate Assignment: Using a CAD system, produce a hard-copy multiview drawing of the problem assigned. Dimension completely.

1

VISIBLE LINE

HIDDEN LINE

SECTION, DIMENSION, AND EXTENSION LINE

CENTER LINE

CUTTING-PLANE LINE

PHANTOM LINE

2

VISIBLE LINE

HIDDEN LINE

SECTION, DIMENSION, AND EXTENSION LINE

CENTER LINE

CUTTING-PLANE LINE

PHANTOM LINE

SHORT-BREAK LINE

3

4

5

6

ALPHABET OF LINES

INSTRUMENTAL DRAWING

DRAWN BY

FILE NO.

DRAWING
1—1

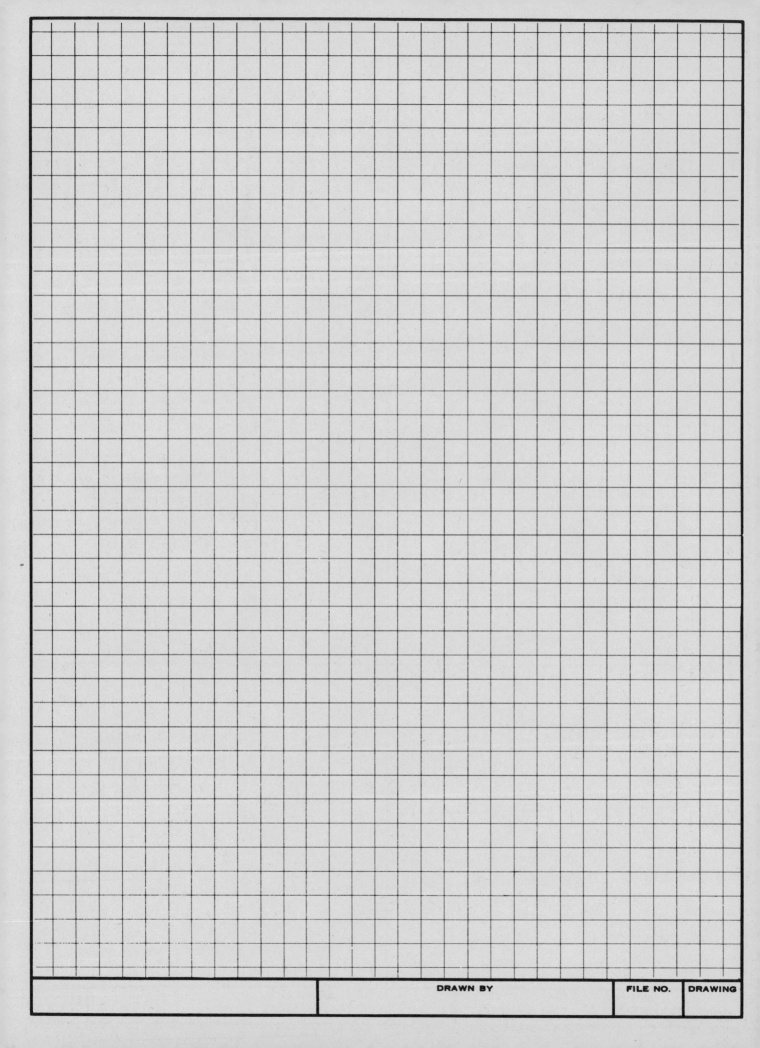

I

A SCALE : 12" = 1'-0   L = ≡≡≡

B SCALE : 6" = 1'-0   L = 11 9/16"

C SCALE : 1" = 10'   L = ≡≡≡

D SCALE : 1" = 20'   L = 97.5'

E SCALE : 1" = 600'   L = 3100.0'

F SCALE : 1mm = 1mm   L = ≡≡≡

G SCALE : 1mm = 10mm   L = 1315.0mm

H SCALE : 1mm = 2mm   L = 286.0mm

J SCALE : ≡≡≡   L = ≡≡≡

K SCALE : ≡≡≡   L = ≡≡≡

L SCALE : ≡≡≡   L = ≡≡≡

Measure or draw the above lengths as indicated and record answers in appropriate spaces.

2

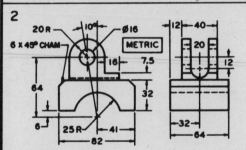

20R   10°   Ø16
6 X 45° CHAM
METRIC
16   7.5
64
32
6
25R   41
82
12 — 40
20
12
32
64

ANCHOR BRACKET

CRS — 6 REQD
SCALE: 1 = 1

Draw the two views.
Omit dimensions.

SCALES AND LAYOUT	DRAWN BY	FILE NO.	DRAWING
INSTRUMENTAL DRAWING			**1-2**

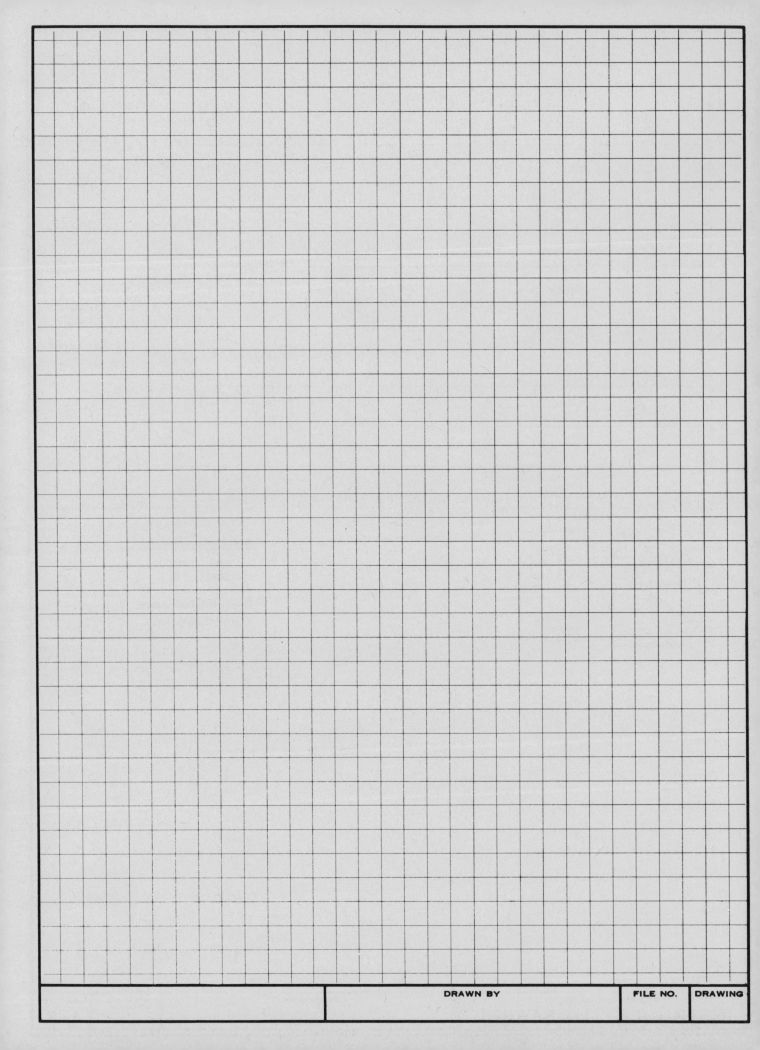

DRAWN BY

FILE NO.

DRAWING

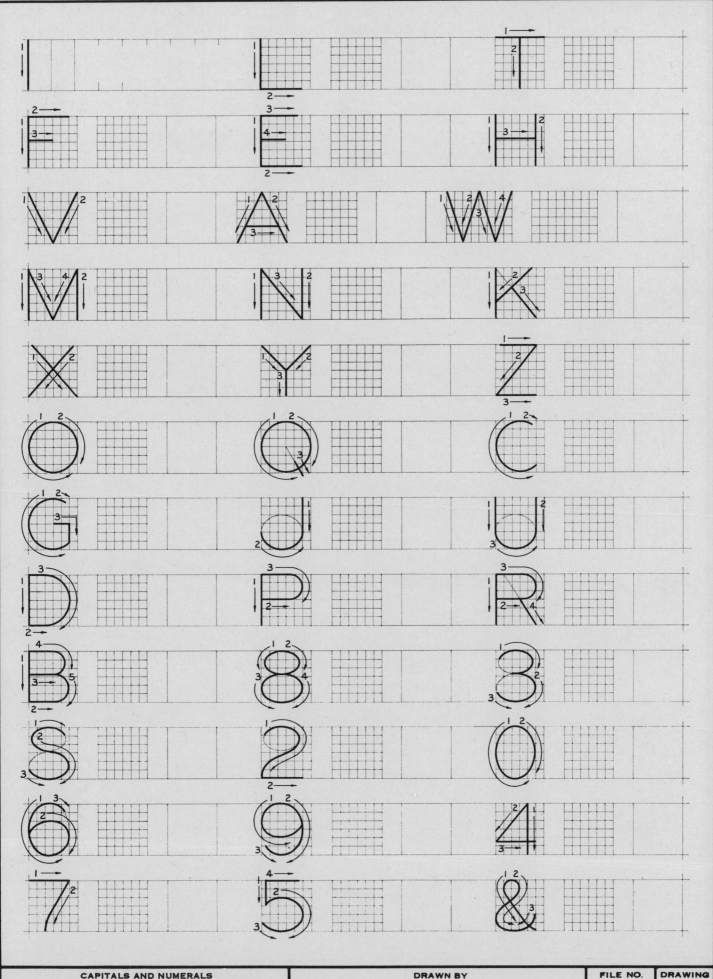

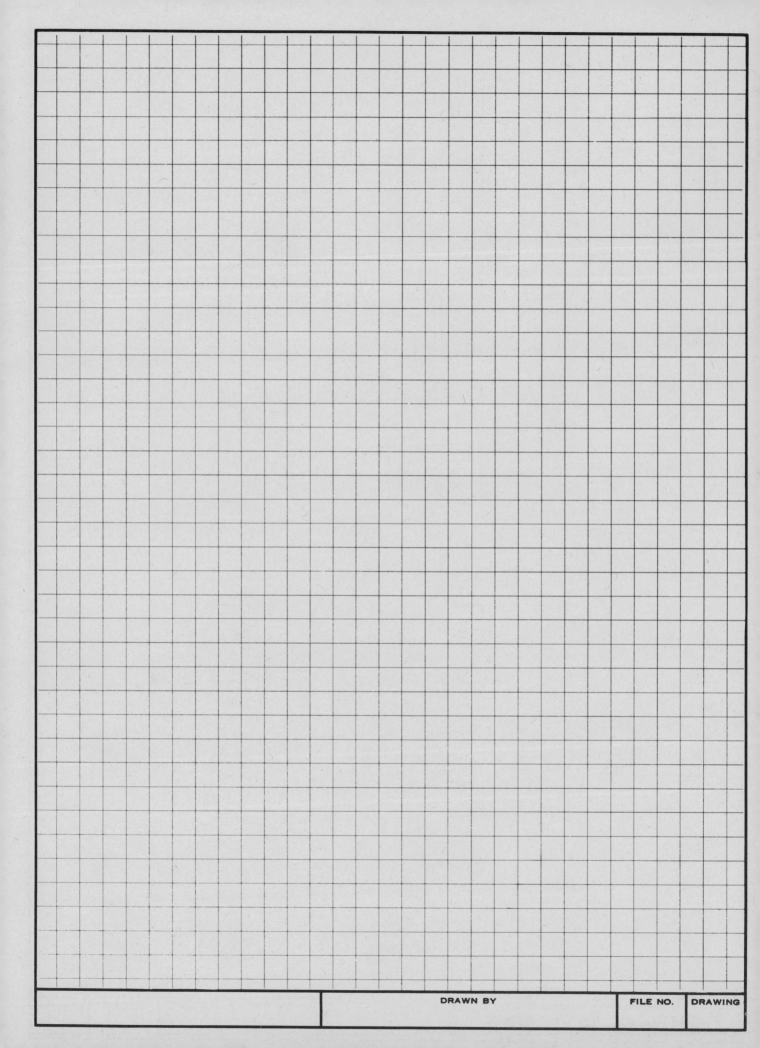

DRAWN BY

FILE NO.

DRAWING

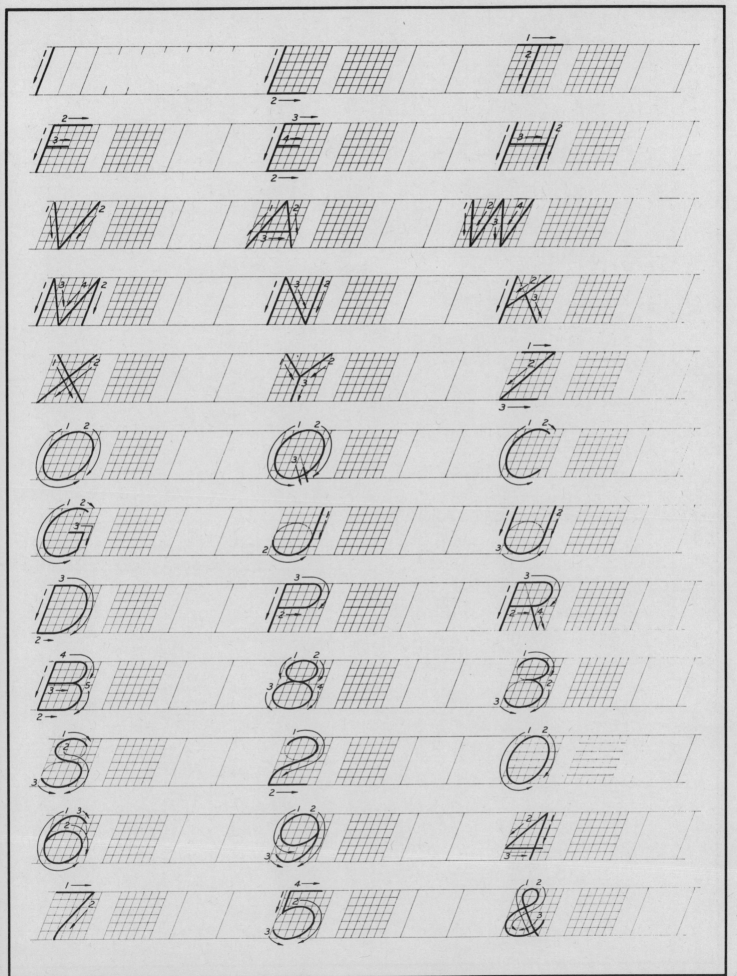

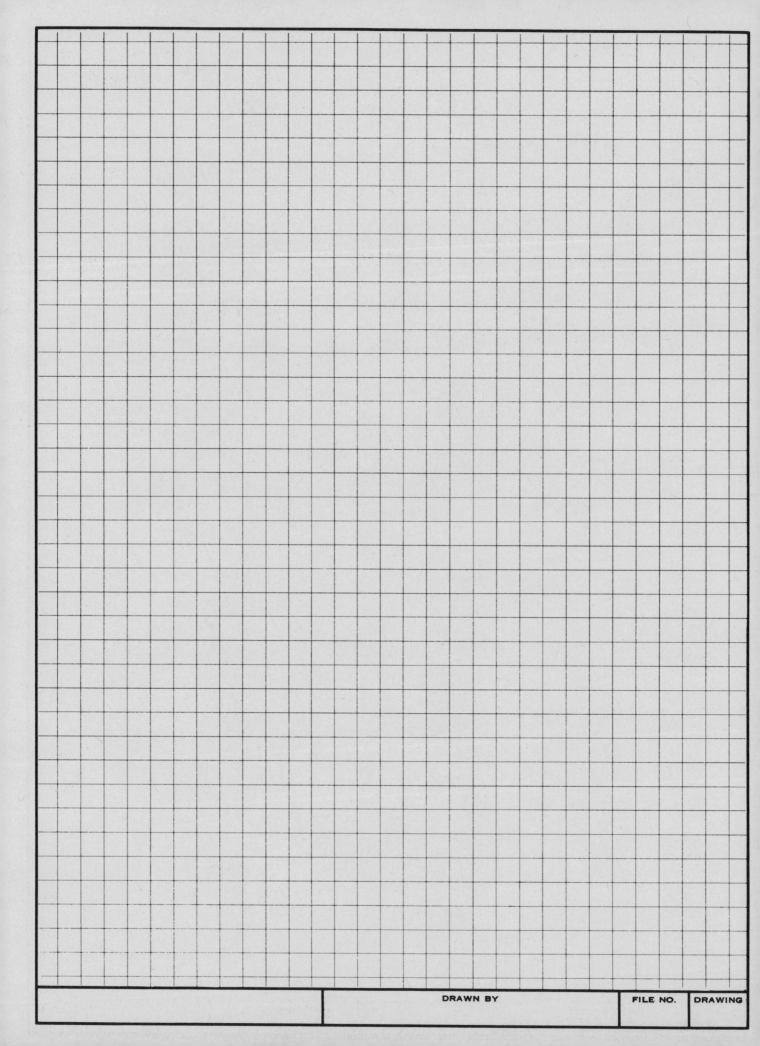

DRAWN BY

FILE NO.

DRAWING

WHILE IT IS TRUE THAT

"PRACTICE MAKES PERFECT," IT

MUST BE UNDERSTOOD THAT

PRACTICE IS NOT ENOUGH, BUT IT

MUST BE ACCOMPANIED BY A CON-

TINUOUS EFFORT TO IMPROVE. EXCEL-

LENT LETTERERS ARE OFTEN NOT GOOD

WRITERS. USE A FAIRLY SOFT PENCIL, AND AL-

WAYS KEEP IT SHARP, ESPECIALLY FOR SMALL

LETTERS. MAKE THE LETTERS CLEAN-CUT AND

DARK-NEVER FUZZY, GRAY, OR INDEFINITE. 1234

$1\frac{1}{2}$ 1.500 $\frac{3}{16}$ 45'-6 32° 15.489 $\frac{13}{64}$ 12"=1'-0 $7\frac{5}{16}$ 12.3 $\frac{1}{2}$ $2\frac{1}{4}$

ONE MUST HAVE A CLEAR MENTAL IMAGE OF THE LETTERS. 234

CAPITALS AND NUMERALS
VERTICAL LETTERING

DRAWN BY

FILE NO.

DRAWING
2-3

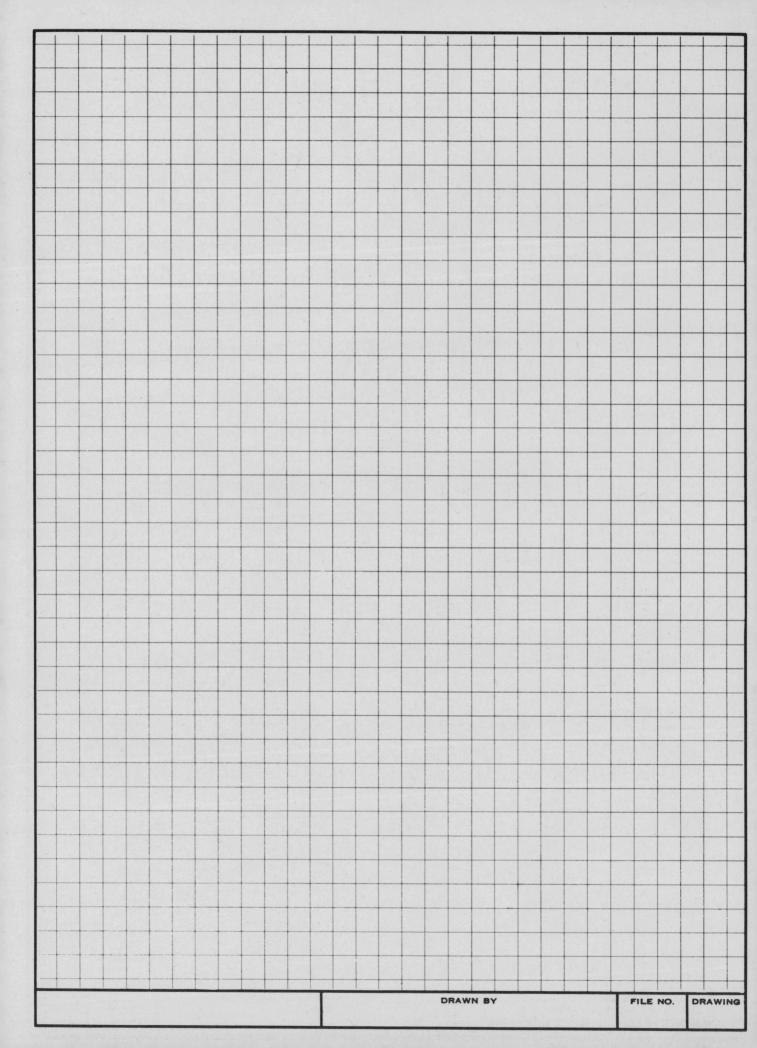

DRAWN BY   FILE NO.   DRAWING

WHILE IT IS TRUE THAT

"PRACTICE MAKES PERFECT," IT

MUST BE UNDERSTOOD THAT

PRACTICE IS NOT ENOUGH, BUT IT

MUST BE ACCOMPANIED BY A CON-

TINUOUS EFFORT TO IMPROVE. EXCEL-

LENT LETTERERS ARE OFTEN NOT GOOD

WRITERS. USE A FAIRLY SOFT PENCIL, AND AL-

WAYS KEEP IT SHARP, ESPECIALLY FOR SMALL

LETTERS. MAKE THE LETTERS CLEAN-CUT AND

DARK-NEVER FUZZY, GRAY, OR INDEFINITE. 1234

$1\frac{1}{2}$ 1.500 $\frac{3}{16}$ 45'-6 32° 15.489 $\frac{13}{64}$ 12"=1'-0 $7\frac{5}{16}$ 12.3 $\frac{1}{2}$ $2\frac{1}{4}$

ONE MUST HAVE A CLEAR MENTAL IMAGE OF THE LETTERS. 234

CAPITALS AND NUMERALS **INCLINED LETTERING**	DRAWN BY	FILE NO.	DRAWING 2–4

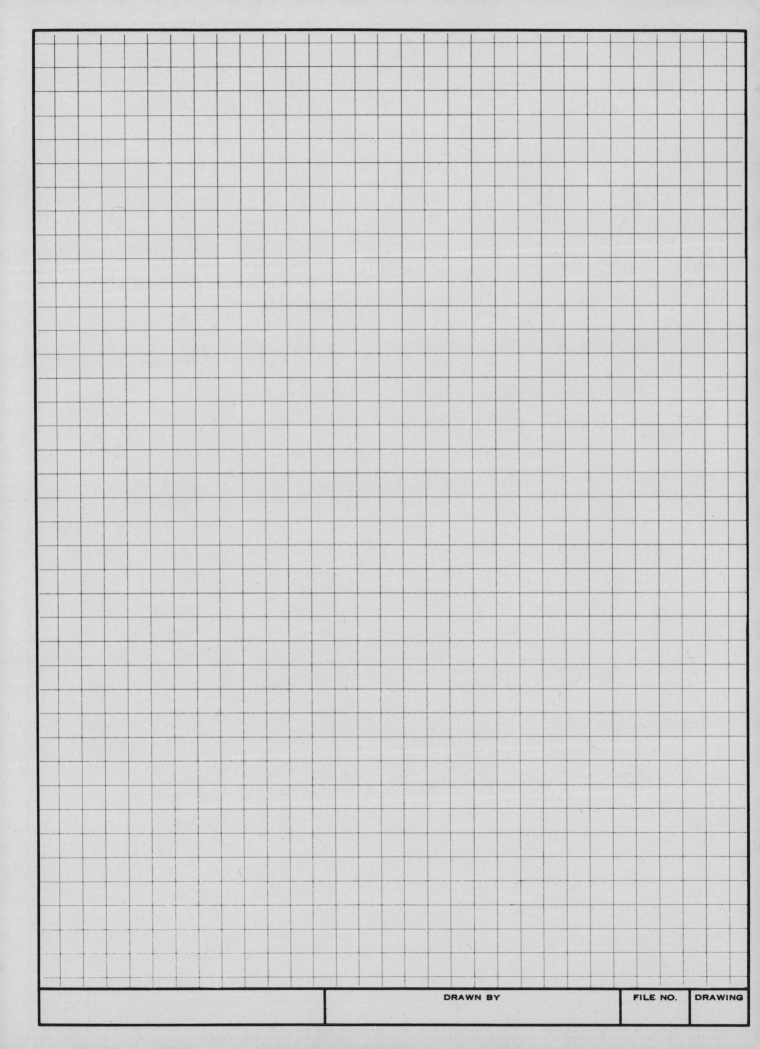

DRAWN BY

FILE NO.

DRAWING

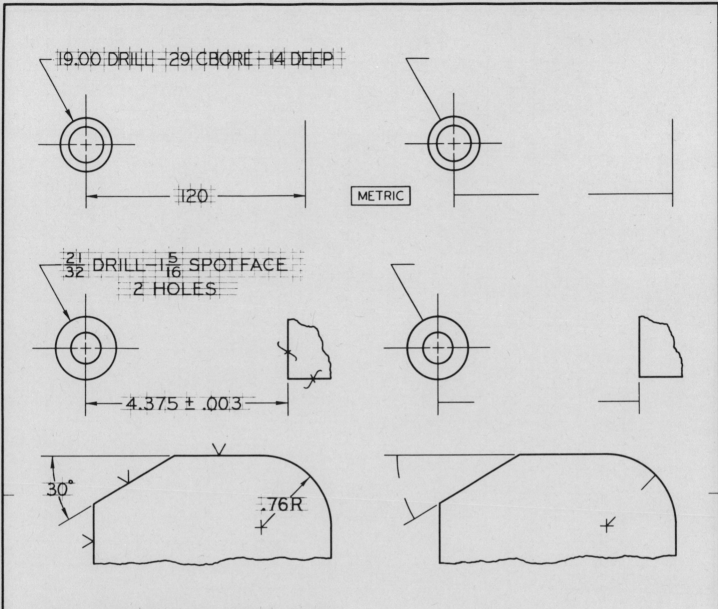

19.00 DRILL - 29 CBORE - 14 DEEP

120

METRIC

$\frac{21}{32}$ DRILL - 1$\frac{5}{16}$ SPOTFACE
2 HOLES

4.375 ± .003

30°

.76R

.06 × 45° CHAMFER BOTH ENDS

FILE FINISH AND POLISH

.562 - .564 REAM - 2 HOLES

M18 × 2.5, 3 HOLES

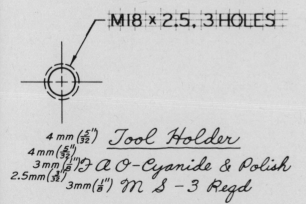

4 mm ($\frac{5}{32}$") *Tool Holder*
4 mm ($\frac{5}{32}$")
3 mm ($\frac{1}{8}$") *F & O - Cyanide & Polish*
2.5 mm ($\frac{3}{32}$")
3 mm ($\frac{1}{8}$") *M S - 3 Reqd*

SHOP NOTES AND DIMENSIONS		DRAWN BY	FILE NO.	DRAWING
VERTICAL LETTERING				**2-5**

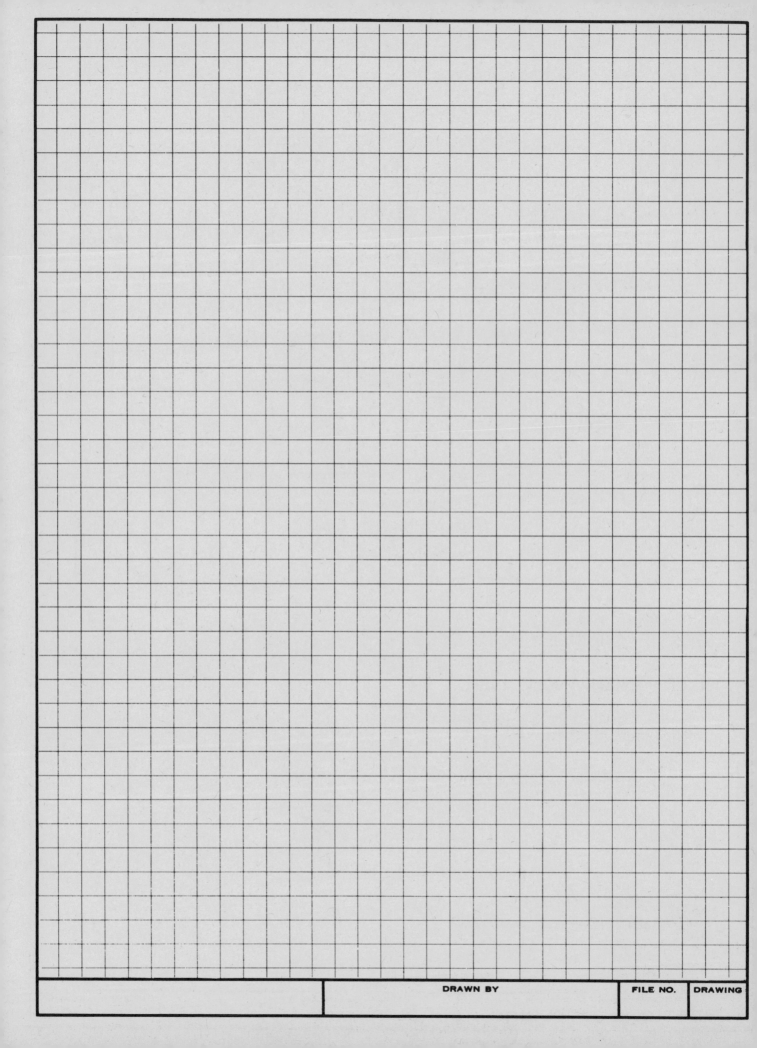

DRAWN BY

FILE NO.

DRAWING

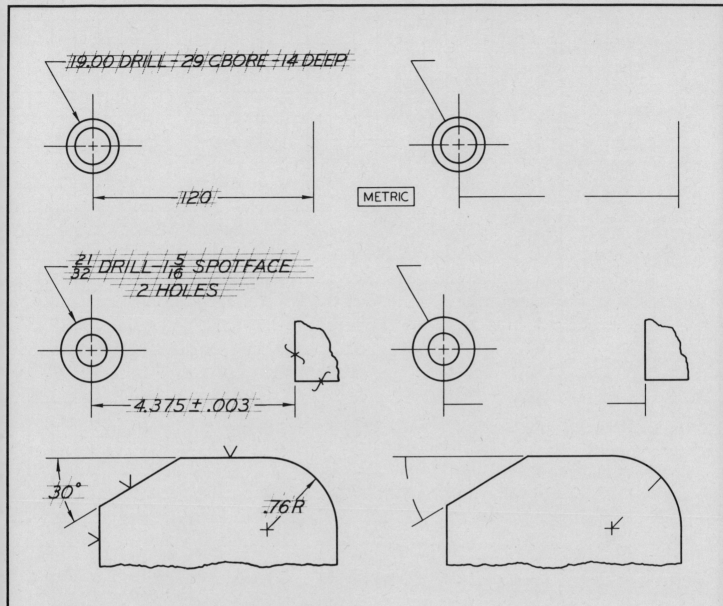

19.00 DRILL - 29 CBORE - 14 DEEP

←————— 120 —————→

METRIC

$\frac{21}{32}$ DRILL - 1 $\frac{5}{16}$ SPOTFACE
2 HOLES

←————— 4.375 ± .003 —————→

30°

.76 R

.06 x 45° CHAMFER BOTH ENDS

FILE FINISH AND POLISH

.562 -.564 REAM - 2 HOLES

M18 x 2.5. 3 HOLES

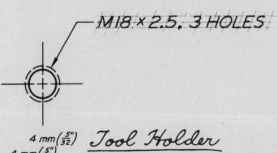

4 mm ($\frac{5''}{32}$)   *Tool Holder*
4 mm ($\frac{5''}{32}$)
3 mm ($\frac{1''}{8}$) *F A O - Cyanide & Polish*
2.5 mm ($\frac{3''}{32}$)
3 mm ($\frac{1''}{8}$) *M S - 3 Reqd*

SHOP NOTES AND DIMENSIONS	DRAWN BY	FILE NO.	DRAWING
INCLINED LETTERING			**2-6**

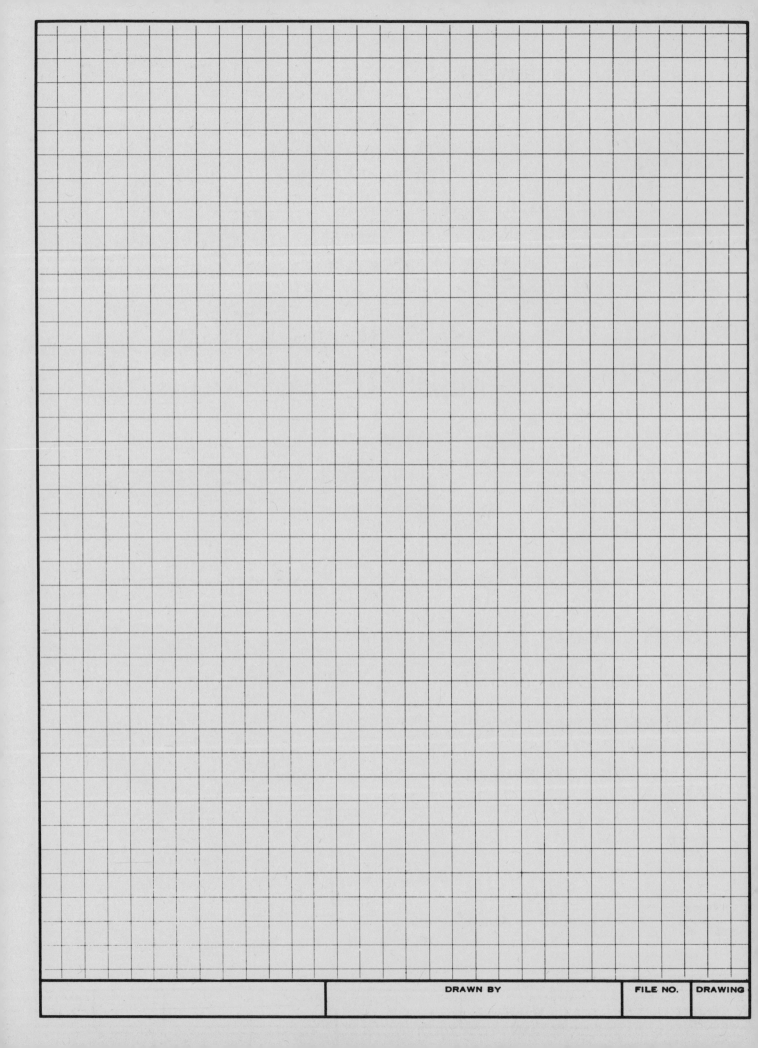

DRAWN BY

FILE NO.

DRAWING

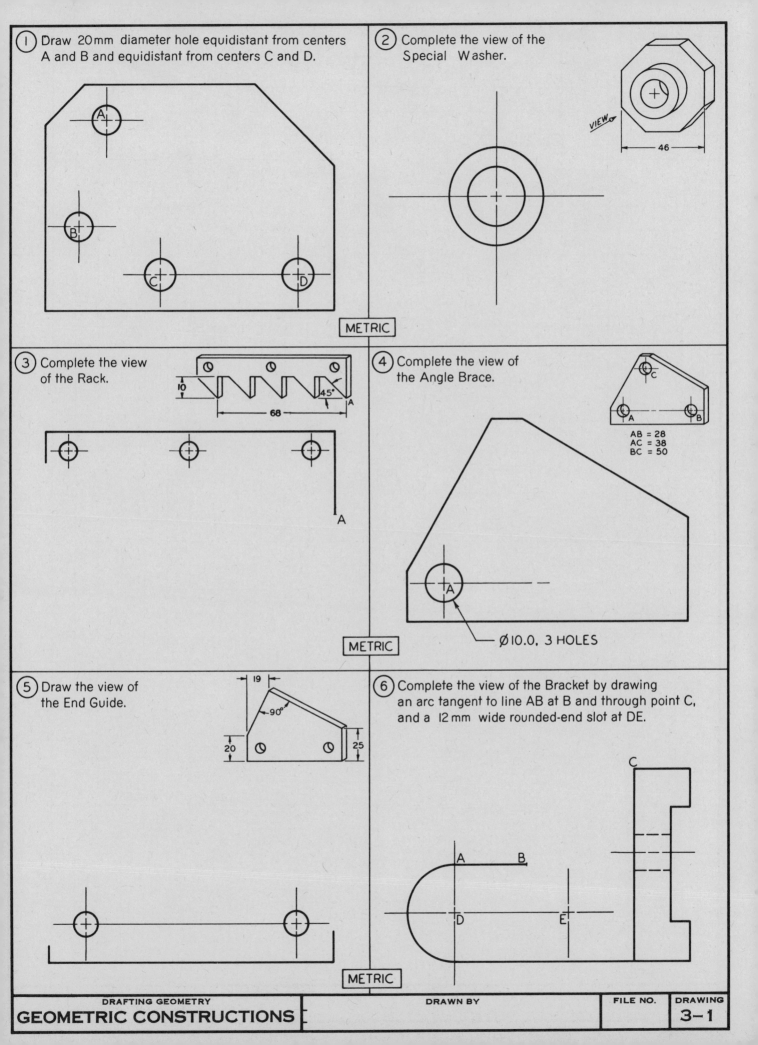

1. Draw 20mm diameter hole equidistant from centers A and B and equidistant from centers C and D.

2. Complete the view of the Special Washer.

VIEW

46

METRIC

3. Complete the view of the Rack.

10

68

45°

A

A

4. Complete the view of the Angle Brace.

C

A

B

AB = 28
AC = 38
BC = 50

A

Ø10.0, 3 HOLES

METRIC

5. Draw the view of the End Guide.

19

90°

20

25

6. Complete the view of the Bracket by drawing an arc tangent to line AB at B and through point C, and a 12 mm wide rounded-end slot at DE.

C

A   B

D   E

METRIC

DRAFTING GEOMETRY
GEOMETRIC CONSTRUCTIONS

DRAWN BY

FILE NO.

DRAWING
3-1

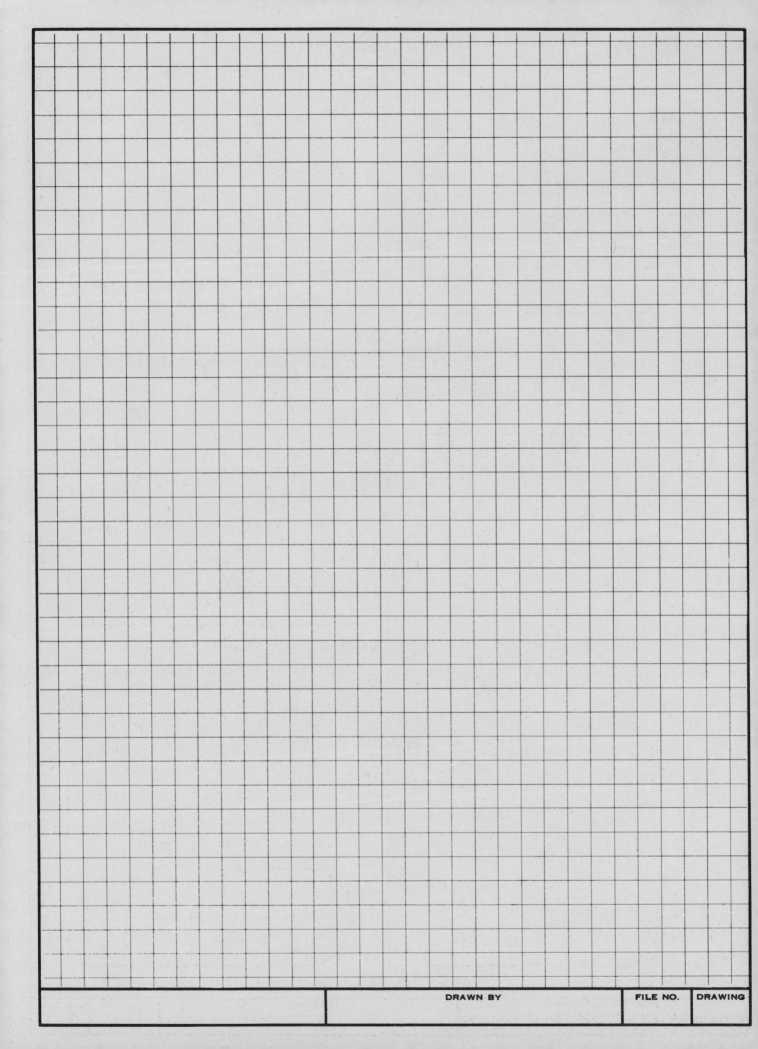

DRAWN BY

FILE NO.

DRAWING

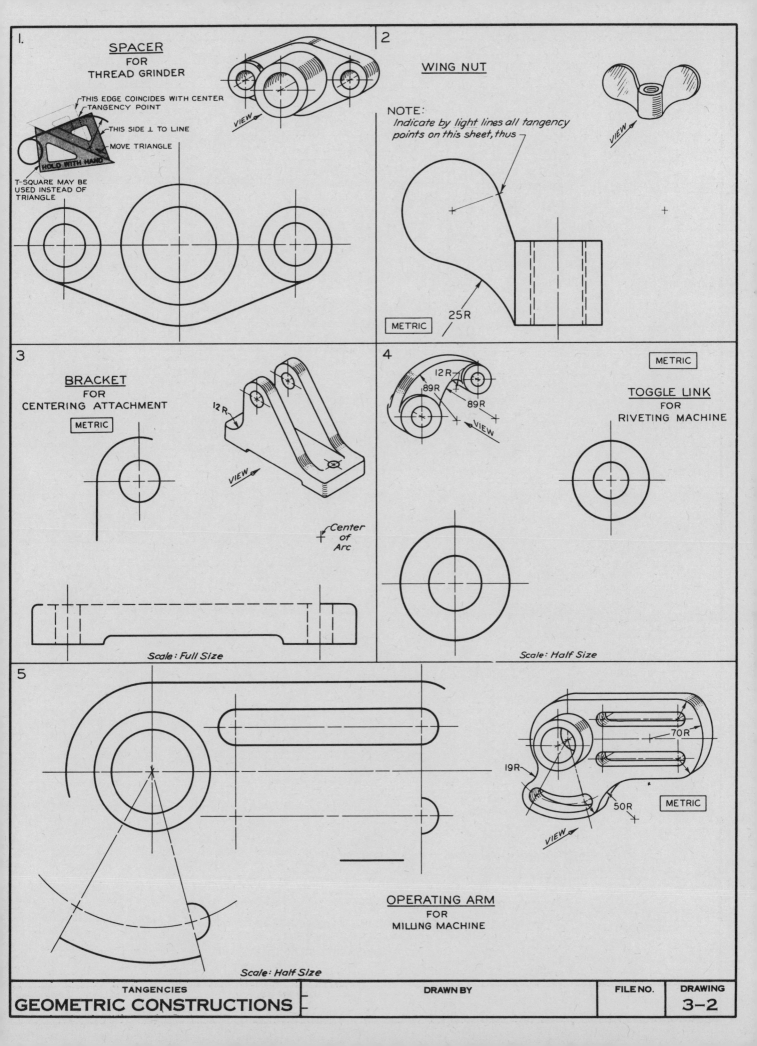

1.

SPACER
FOR
THREAD GRINDER

THIS EDGE COINCIDES WITH CENTER TANGENCY POINT

THIS SIDE ⊥ TO LINE

MOVE TRIANGLE

HOLD WITH HAND

T-SQUARE MAY BE USED INSTEAD OF TRIANGLE

VIEW

2

WING NUT

NOTE:
Indicate by light lines all tangency points on this sheet, thus

VIEW

25R

METRIC

3

BRACKET
FOR
CENTERING ATTACHMENT

METRIC

12 R

VIEW

Center
of
Arc

Scale: Full Size

4

METRIC

12 R

89R

89R

VIEW

TOGGLE LINK
FOR
RIVETING MACHINE

Scale: Half Size

5

70R

19R

50R

METRIC

VIEW

OPERATING ARM
FOR
MILLING MACHINE

Scale: Half Size

TANGENCIES
GEOMETRIC CONSTRUCTIONS

DRAWN BY

FILE NO.

DRAWING
3-2

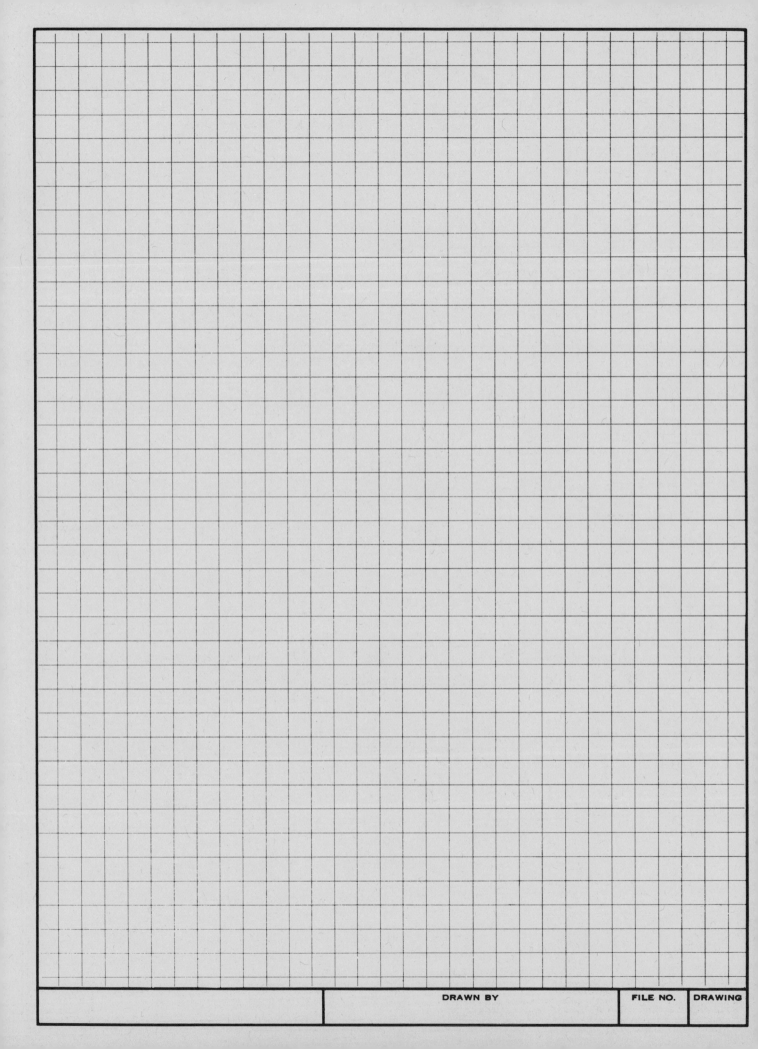

DRAWN BY · FILE NO. · DRAWING

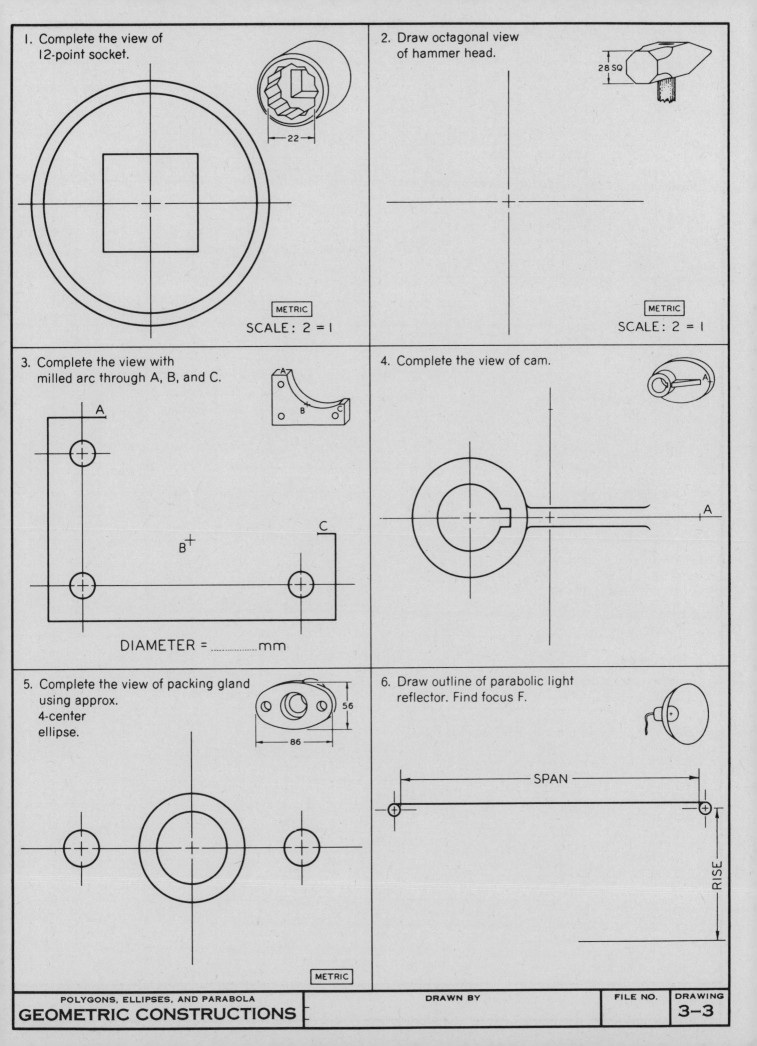

1. Complete the view of 12-point socket.

METRIC
SCALE: 2 = 1

—22—

2. Draw octagonal view of hammer head.

28 SQ

METRIC
SCALE: 2 = 1

3. Complete the view with milled arc through A, B, and C.

A

A

B+

C

DIAMETER = ... mm

4. Complete the view of cam.

A

A

5. Complete the view of packing gland using approx. 4-center ellipse.

56

86

METRIC

6. Draw outline of parabolic light reflector. Find focus F.

SPAN

RISE

POLYGONS, ELLIPSES, AND PARABOLA

GEOMETRIC CONSTRUCTIONS

DRAWN BY

FILE NO.

DRAWING
3-3

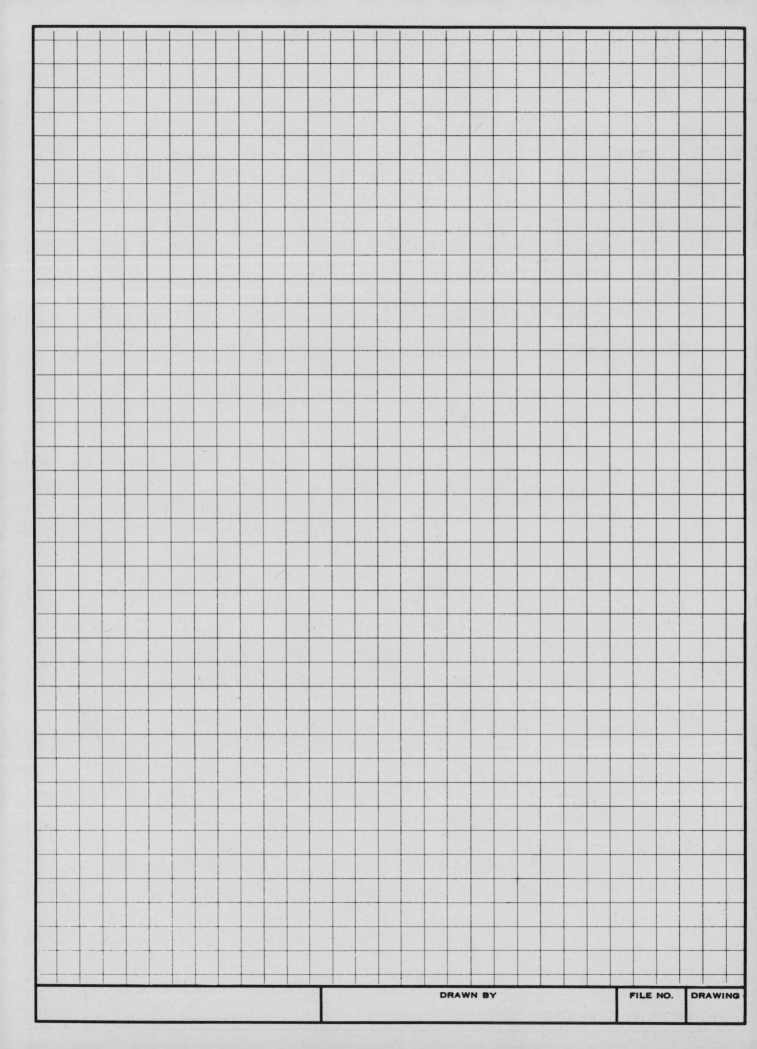

DRAWN BY

FILE NO.

DRAWING

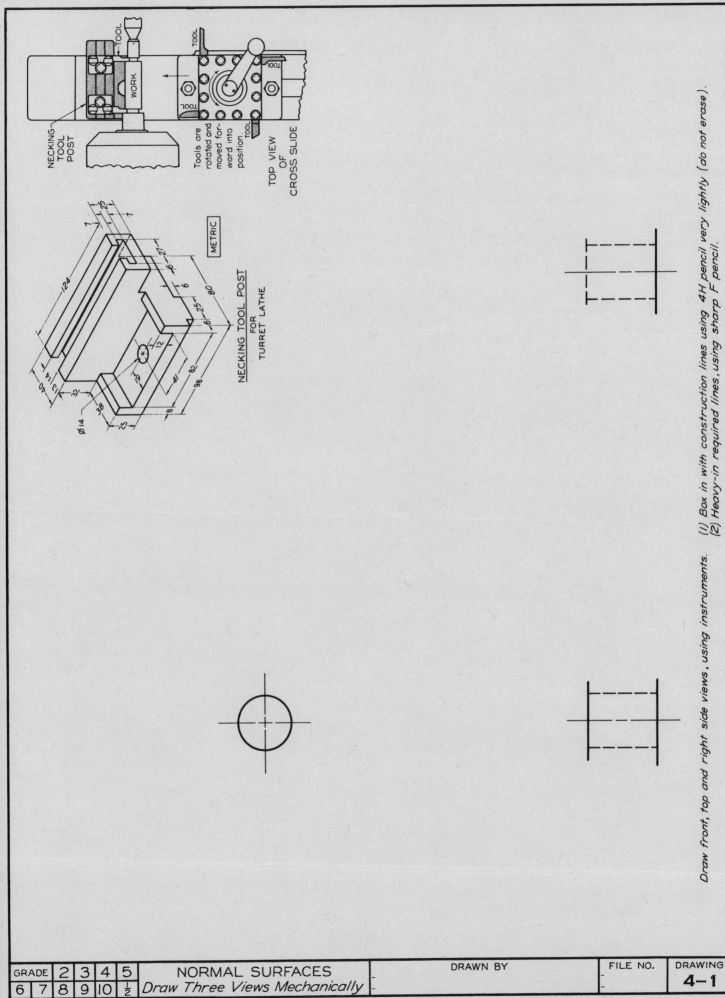

NECKING TOOL POST

Tools are rotated and moved forward into position.

TOP VIEW OF CROSS SLIDE

METRIC

NECKING TOOL POST FOR TURRET LATHE

Draw front, top and right side views, using instruments.   (1) Box in with construction lines using 4H pencil very lightly (do not erase).
(2) Heavy-in required lines, using sharp F pencil.

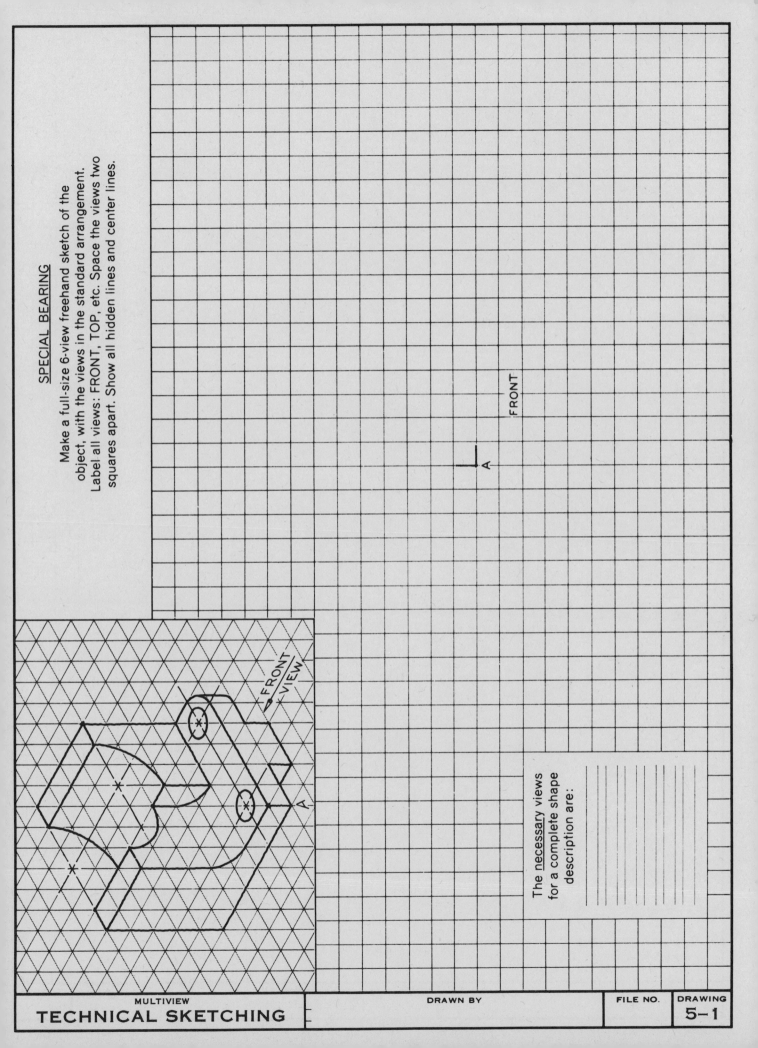

SPECIAL BEARING

Make a full-size 6-view freehand sketch of the object, with the views in the standard arrangement. Label all views: FRONT, TOP, etc. Space the views two squares apart. Show all hidden lines and center lines.

FRONT

FRONT VIEW

A

The necessary views for a complete shape description are:

MULTIVIEW

TECHNICAL SKETCHING

DRAWN BY

FILE NO.

DRAWING

5—1

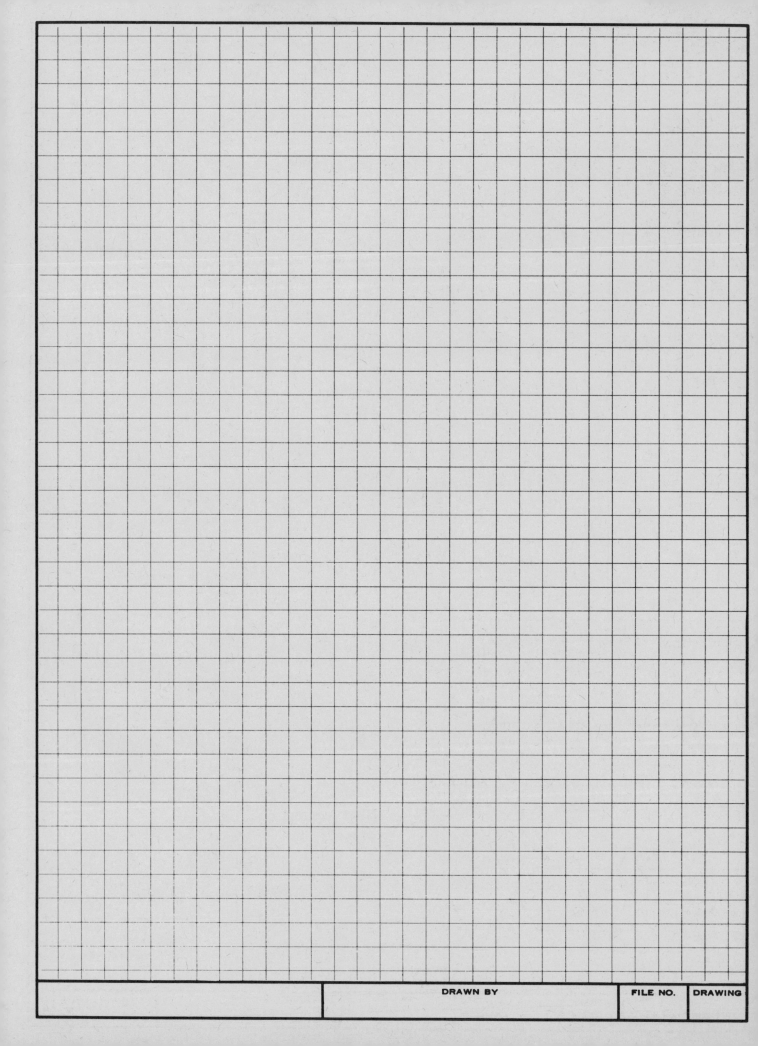

DRAWN BY   FILE NO.   DRAWING

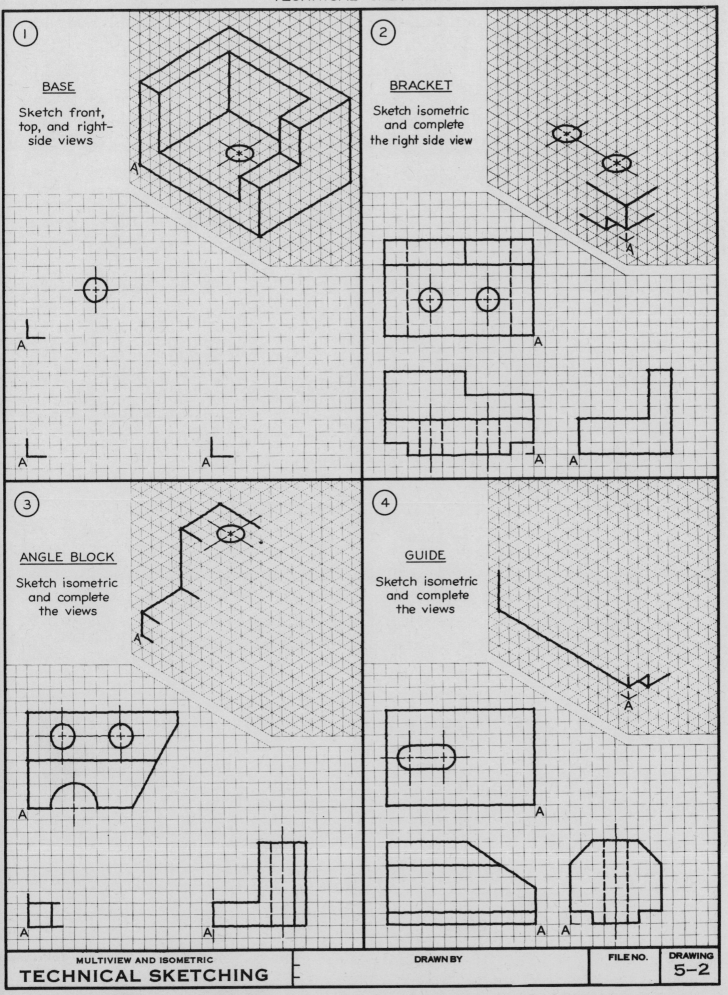

1

<u>BASE</u>

Sketch front, top, and right-side views

2

<u>BRACKET</u>

Sketch isometric and complete the right side view

3

<u>ANGLE BLOCK</u>

Sketch isometric and complete the views

4

<u>GUIDE</u>

Sketch isometric and complete the views

MULTIVIEW AND ISOMETRIC	DRAWN BY	FILE NO.	DRAWING
TECHNICAL SKETCHING			5-2

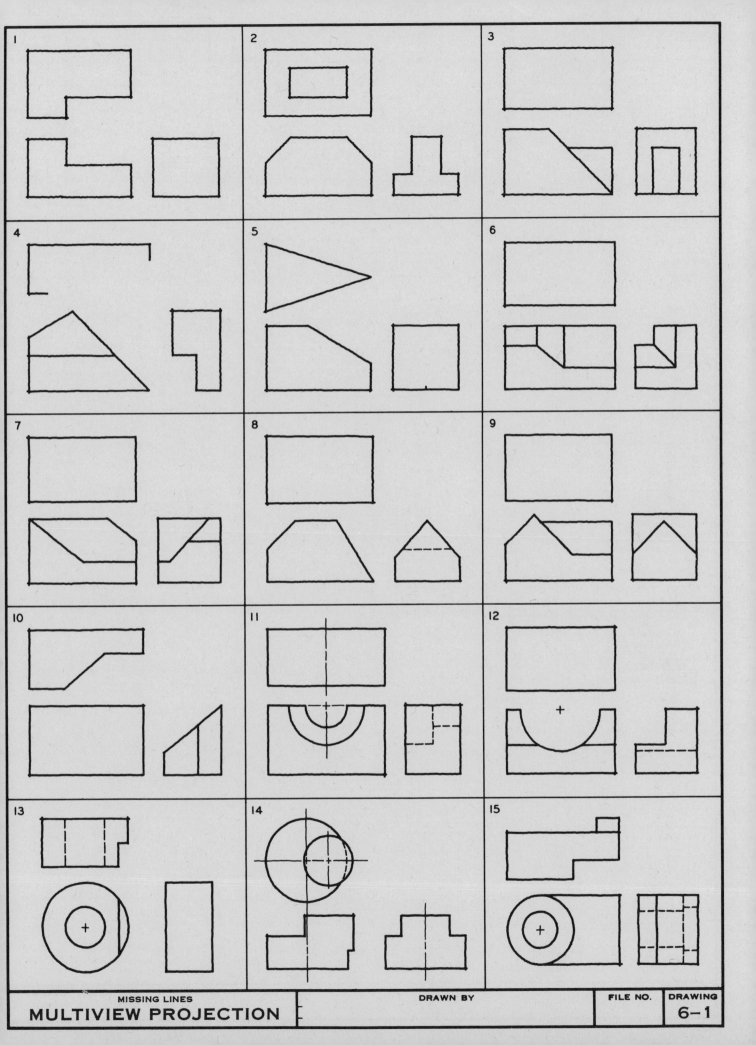

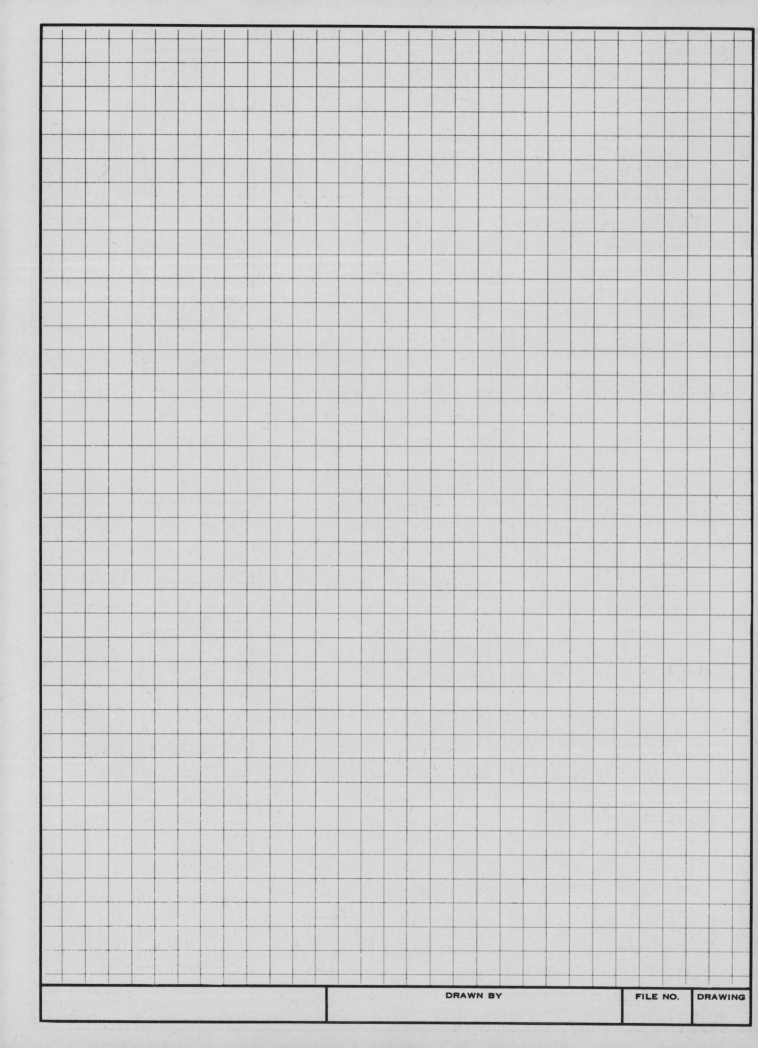

DRAWN BY

FILE NO.

DRAWING

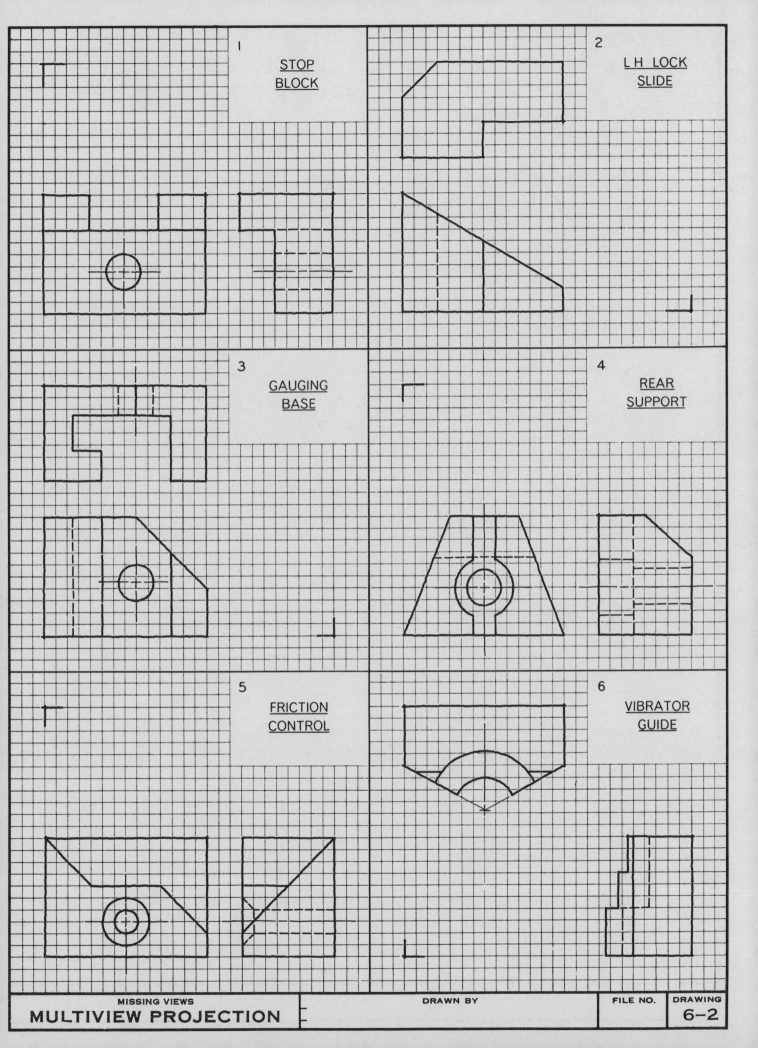

1 STOP BLOCK

2 L H LOCK SLIDE

3 GAUGING BASE

4 REAR SUPPORT

5 FRICTION CONTROL

6 VIBRATOR GUIDE

MISSING VIEWS

MULTIVIEW PROJECTION

DRAWN BY

FILE NO.

DRAWING 6-2

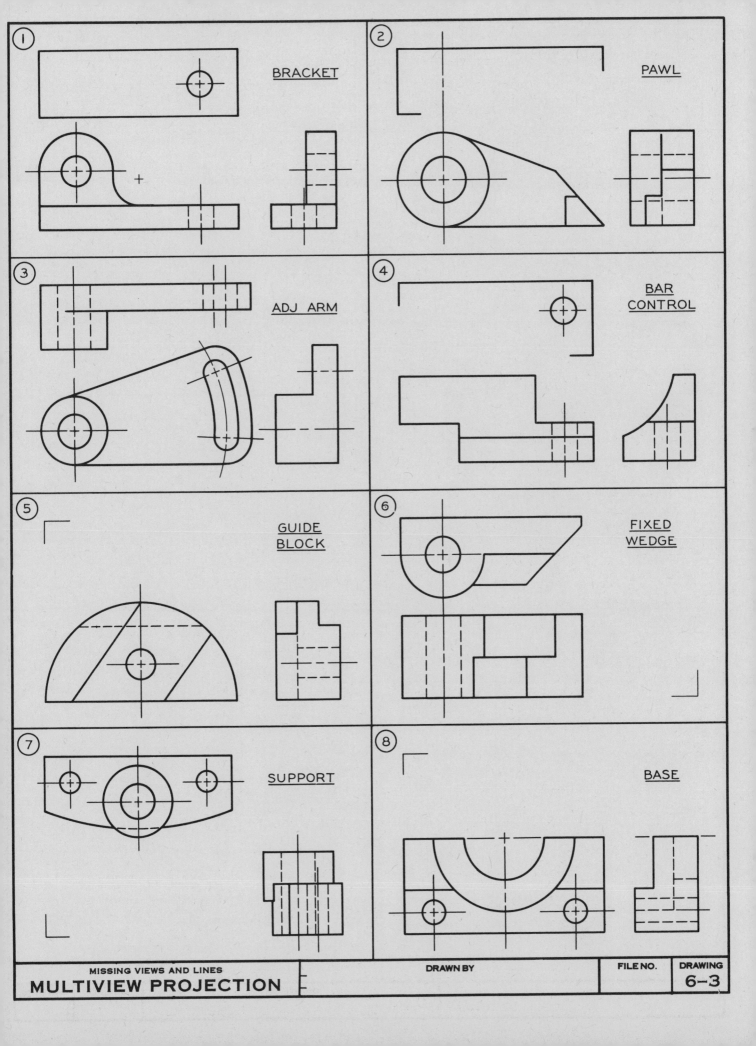

1. BRACKET

2. PAWL

3. ADJ ARM

4. BAR CONTROL

5. GUIDE BLOCK

6. FIXED WEDGE

7. SUPPORT

8. BASE

| MISSING VIEWS AND LINES | DRAWN BY | FILE NO. | DRAWING |
| MULTIVIEW PROJECTION | | | 6-3 |

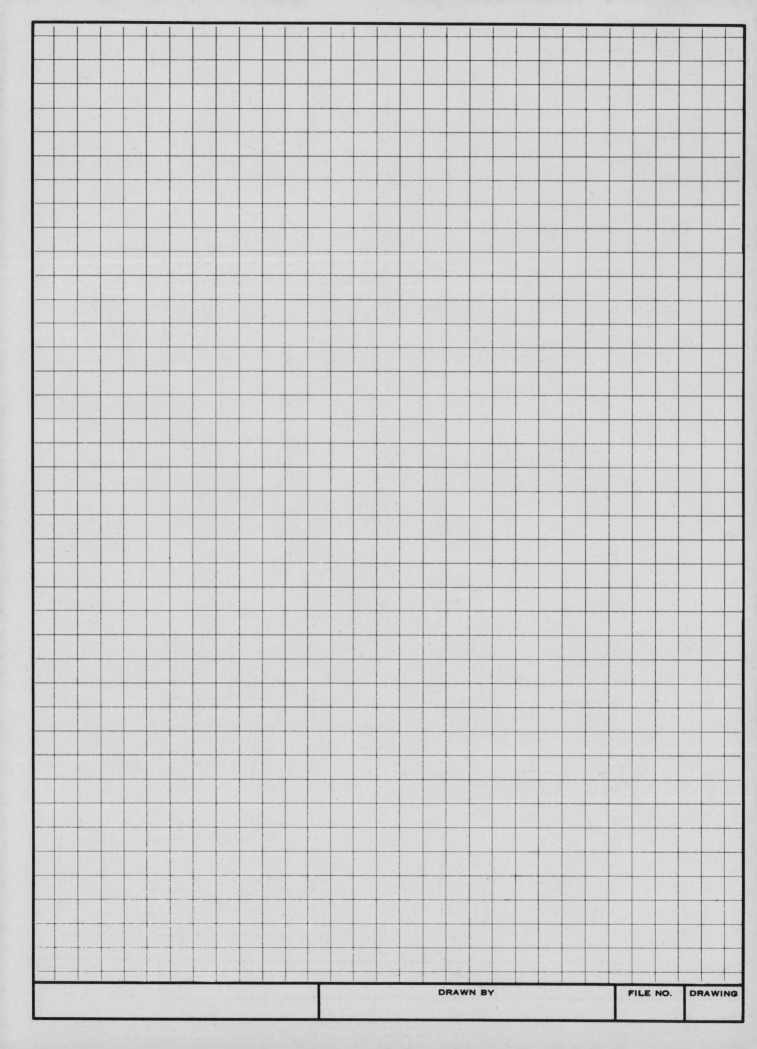

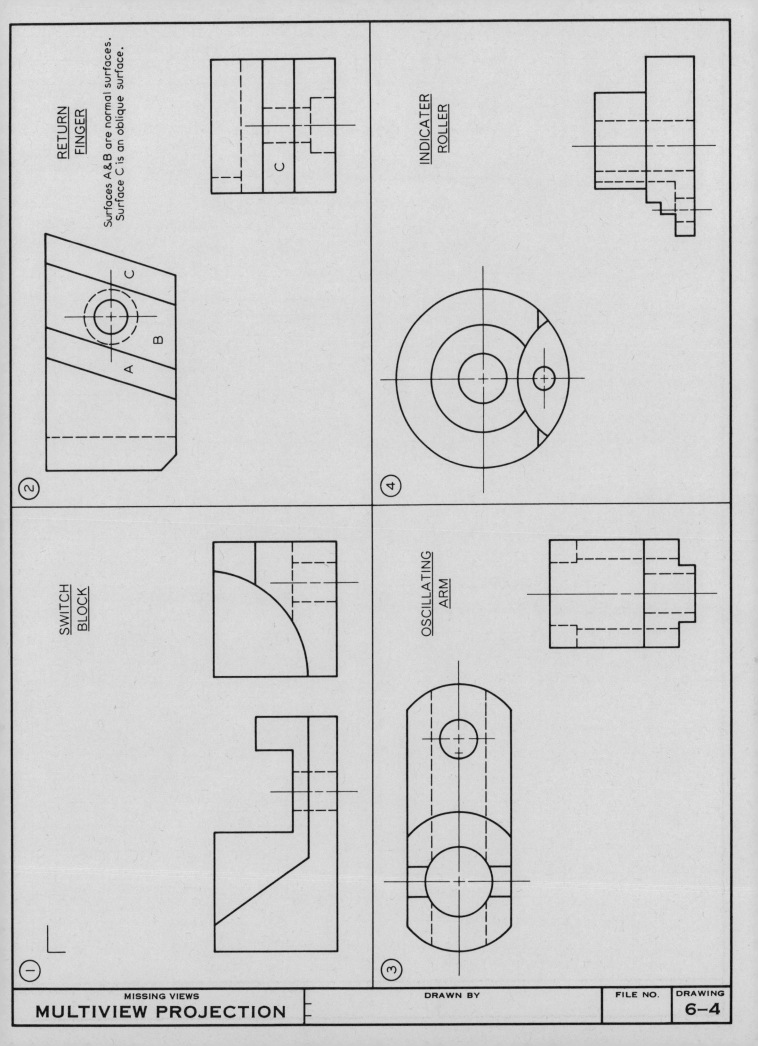

RETURN
FINGER

Surfaces A & B are normal surfaces.
Surface C is an oblique surface.

C

②

INDICATER
ROLLER

④

SWITCH
BLOCK

OSCILLATING
ARM

①

③

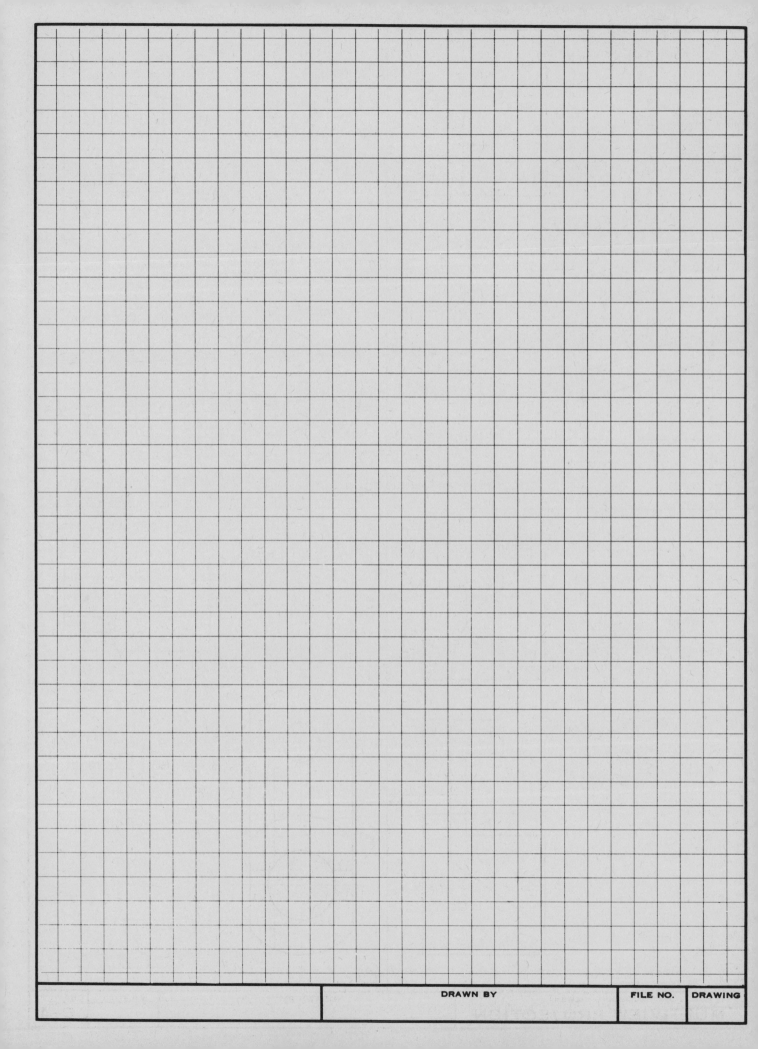

DRAWN BY

FILE NO.

DRAWING

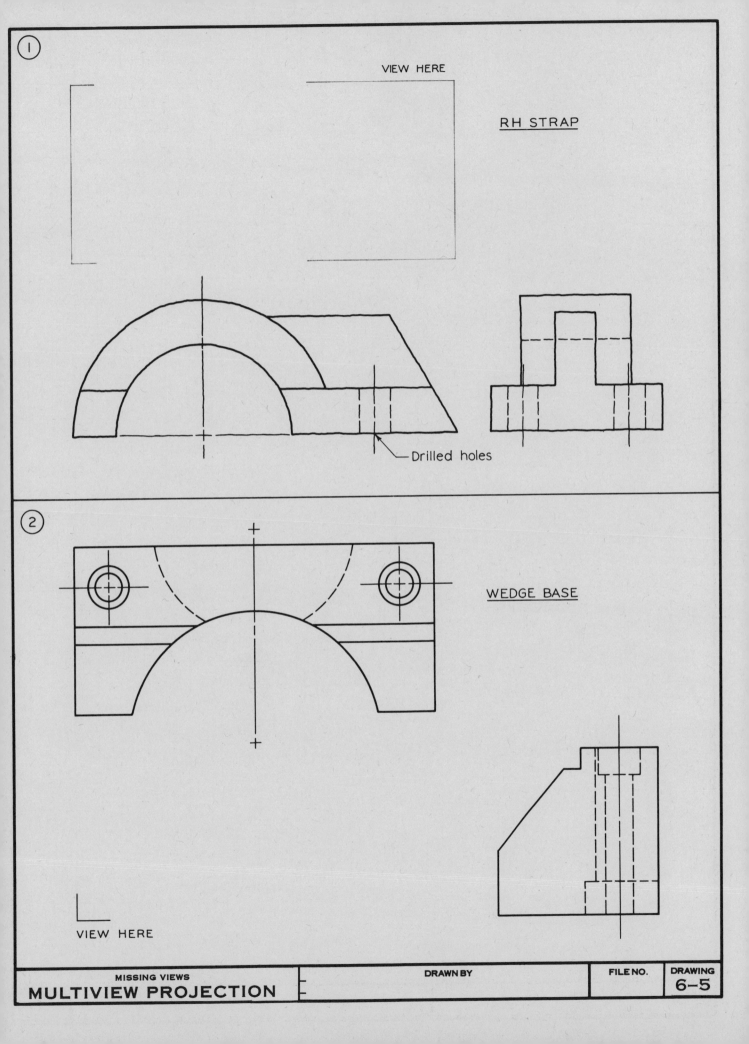

1

VIEW HERE

RH STRAP

Drilled holes

2

WEDGE BASE

VIEW HERE

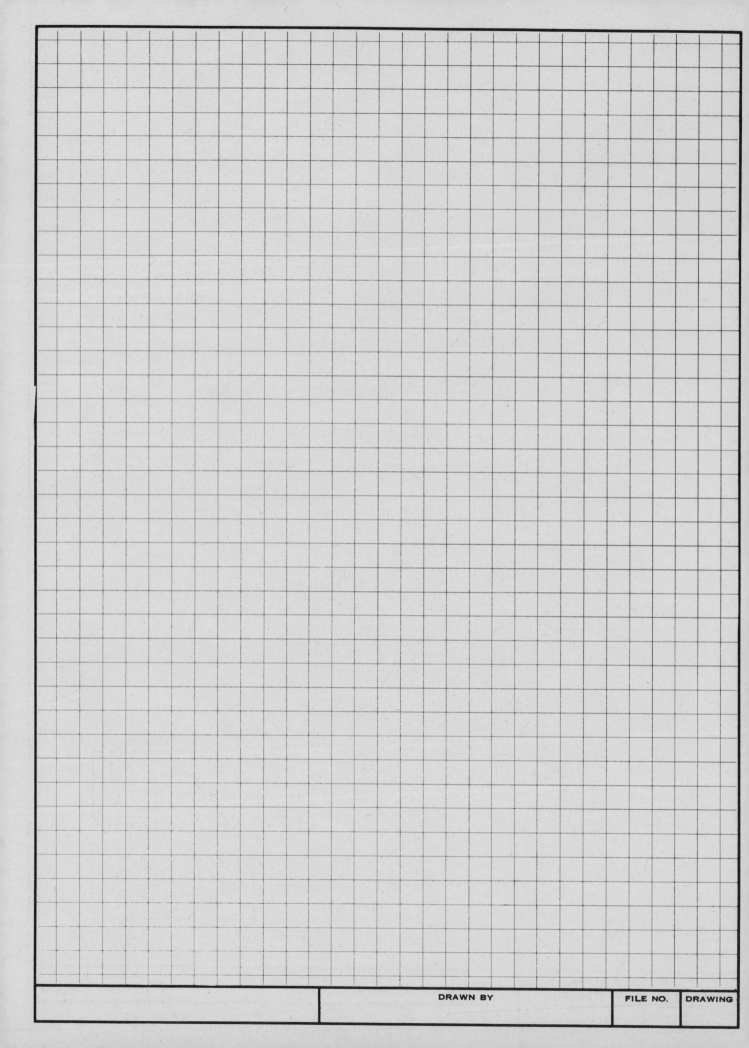

DRAWN BY

FILE NO.

DRAWING

Complete the table of TERMS by entering the letter identifiers of the matching descriptions.

TERMS

TERMS	
CURSOR	
DIGITIZER TABLET	
GRAPHIC PRIMITIVE	
PIXEL	
RESOLUTION	
RASTER DISPLAY	
RAM	
DEBUG	
HARD COPY	
ANALOG	
DIGITAL	
CAE	
COMMAND	
PLOTTER	
HARD DISK	
VECTOR	
COORDINATE SYSTEM	
MENU	
BIT	
MOUSE	
SOFTWARE	
JOY STICK	
WINDOW	
TRANSFORM	
BYTE	
LIGHTPEN	

Descriptions

A Handheld pointing device for pick and coordinate entry

B Computer program to perform specific tasks

C Counts in discrete steps or digits

D Smallest unit of digital information

E Collection of commands for selection

F Device to convert analog picture to coordinate digital data

G Fundamental drawing entity

H Picture element dot in a display grid

I Random Access Memory - volatile physical memory

J Continuous measurements without steps

K Computer assisted engineering

L Group of 8 bits commonly used to represent a character

M Paper printout

N Hand controlled lever used as input device

O Smallest spacing between CRT display elements

P Convert an image into a proper display format

Q Directed line segment with magnitude

R Flicker-free scanned CRT surface

S A bounded rectangular area on screen

T A visual tracking symbol

U Handheld photosensitive input device

V Control signal

W Correct errors

X Non-volatile external storage device

Y Hard copy device for vector drawing

Z Common reference system for spatial relationships

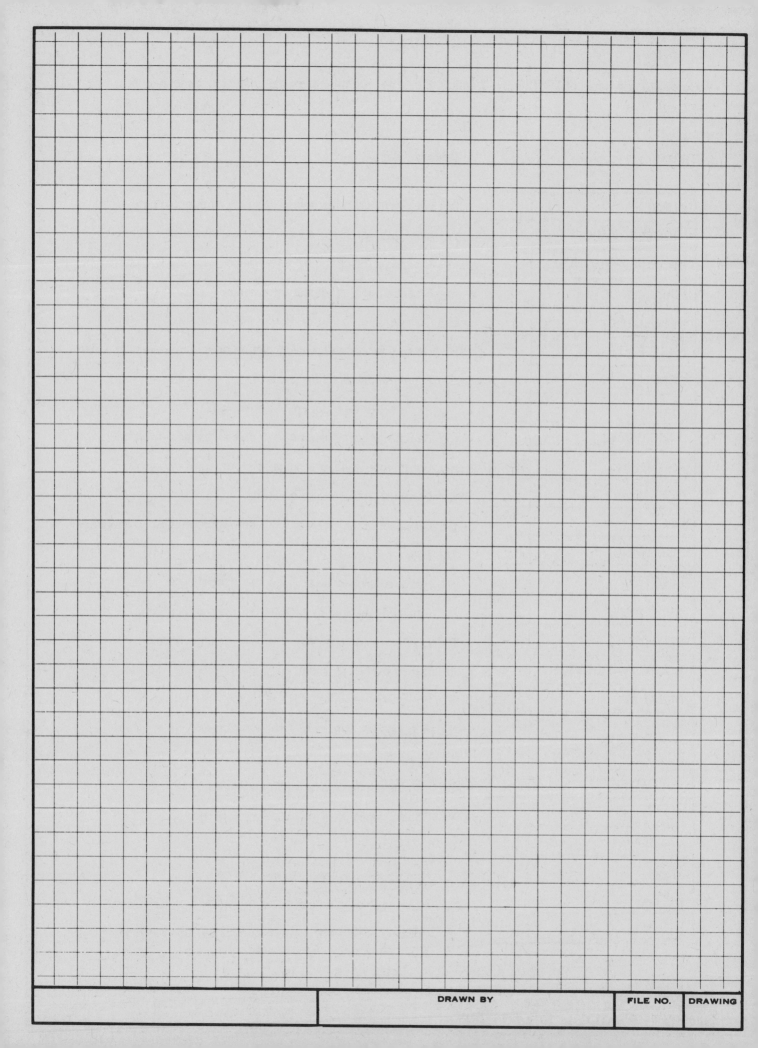

DRAWN BY

FILE NO.

DRAWING

① Complete the table by defining X and Y coordinates of the given points.

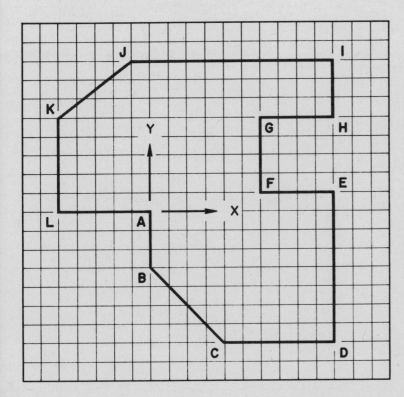

Point	Coordinate	
	X	Y
A		
B		
C		
D		
E		
F		
G		
H		
I		
J		
K		
L		

② Plot the given points on the grid and draw the view.

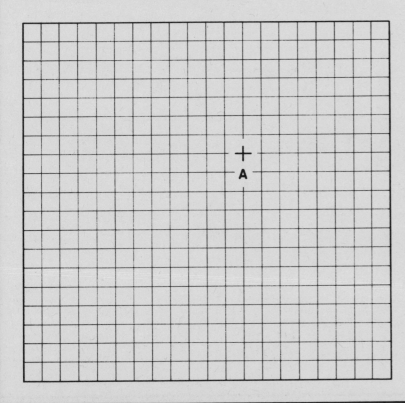

Point	Coordinate	
	X	Y
A	0	0
B	0	4
C	-5	4
D	-5	-2
E	-10	-6
F	-10	-10
G	-2	-10
H	6	-4
I	6	3
J	3	3
K	3	-2
L	0	-4

TWO-DIMENSIONAL COORDINATE PLOT

COMPUTER-AIDED DRAFTING

DRAWN BY

FILE NO.

DRAWING

7-2

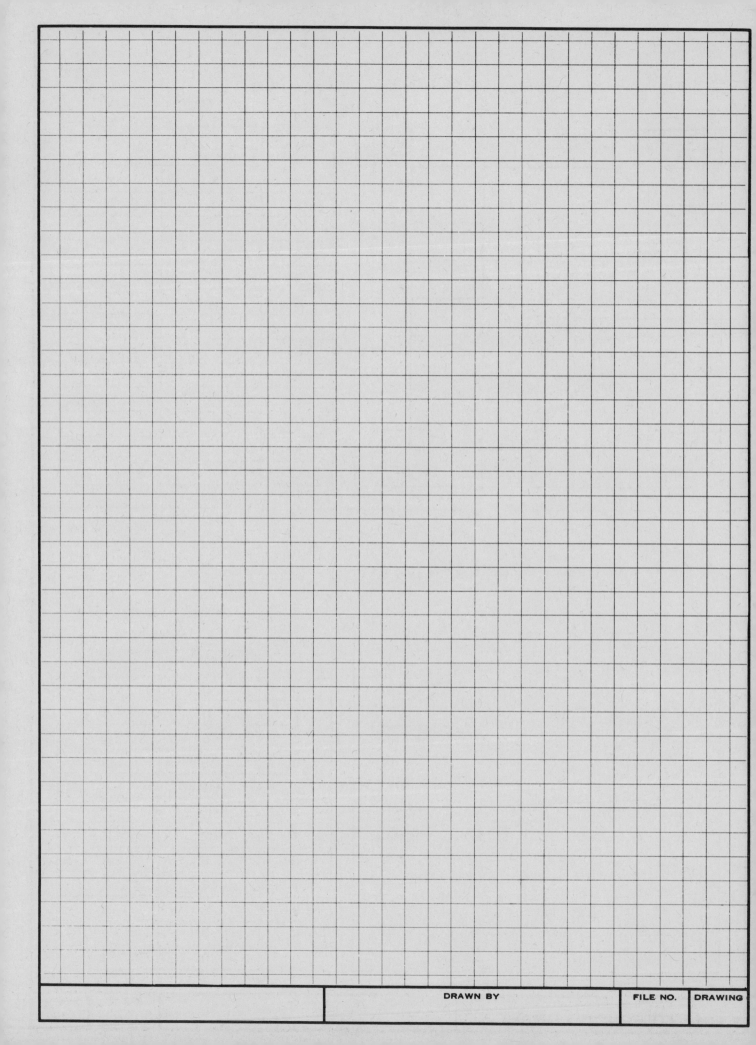

DRAWN BY   FILE NO.   DRAWING

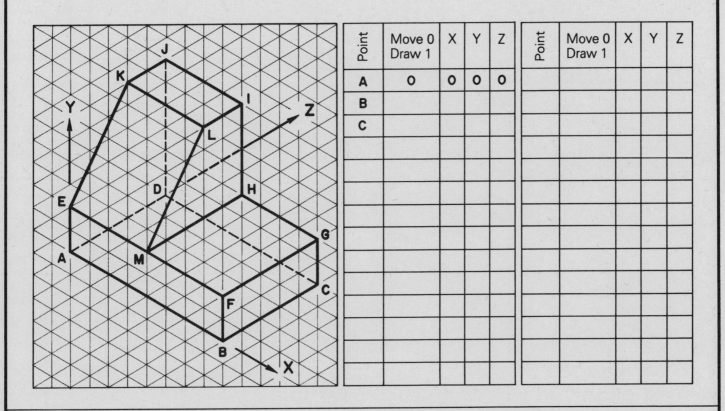

Point	Move 0 Draw 1	X	Y	Z	Point	Move 0 Draw 1	X	Y	Z
A	0	0	0	0					
B									
C									

② Draw the object based on the data given in the table.

Point	Move 0 Draw 1	X	Y	Z	Point	Move 0 Draw 1	X	Y	Z
A	0	0	0	0	G	1	-3	-4	0
B	1	0	5	0	J	1	-3	-4	5
C	1	0	5	5	L	0	-3	-1	0
D	1	0	0	5	B	1	0	5	0
E	1	6	0	5	K	0	-3	-1	5
F	1	6	0	0	C	1	0	5	5
A	1	0	0	0	H	0	0	-4	0
D	1	0	0	5	F	1	6	0	0
G	0	-3	-4	0	I	0	0	-4	5
L	1	-3	-1	0	E	1	6	0	5
K	1	-3	-1	5					
J	1	-3	-4	5					
I	1	0	-4	5					
H	1	0	-4	0					

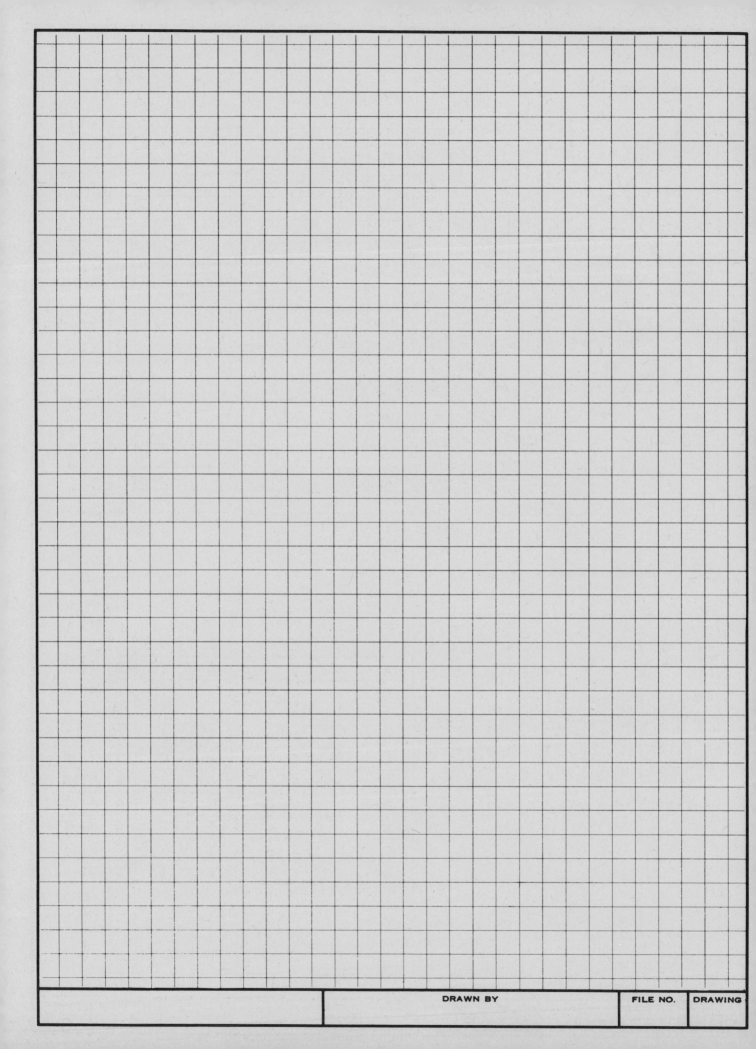

DRAWN BY

FILE NO.

DRAWING

Complete the table by entering the Menu Selections used for generating the drawing.

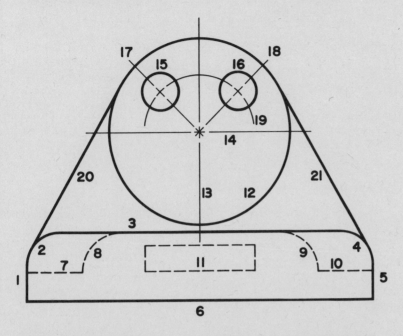

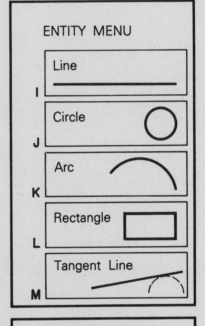

LINE TYPE MENU

A — Visible ————————

B — Hidden – – – – – – – –

C — Center —— — — ——

ENTITY MENU

I — Line ————————

J — Circle

K — Arc

L — Rectangle

M — Tangent Line

CONSTRUCTION MENU

P — From point to point

Q — Around center with radius

R — Around center with radius and angle

S — With height and width

T — From circle to arc

Entity	Line type menu selection	Entity menu selection	Construction menu selection
1			
2			
3			
4			
5			
6			
7			
8			
9			
10			
11			
12			
13			
14			
15			
16			
17			
18			
19			
20			
21			

DRAWN BY

FILE NO.

DRAWING

Description of VIEW COORDINATES

VIEW COORDINATES are the coordinate values of the object as assigned with respect to the computer screen, with X , Y and Z axes positioned as shown below . The coordinates remain the same irrespective of the view selected on the screen .

Axis	Position	Positive Direction
X	Horizontal	To the right
Y	Vertical	Toward the top
Z	Perpendicular to the screen	Outward from the screen

Description of WORLD COORDINATES

WORLD COORDINATES are the coordinate values of the object as assigned with respect to the axes of the object . The X , Y and Z axes are positioned as shown, such that for the top view the X axis is horizontal to the right, the Y axis is vertical to the top and the Z axis is perpendicular to the screen positioned outwards . The coordinates in relation to the screen change according to the view selected on the screen.

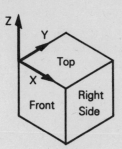

Complete the tables by entering the VIEW and WORLD COORDINATES of the given points of the object .

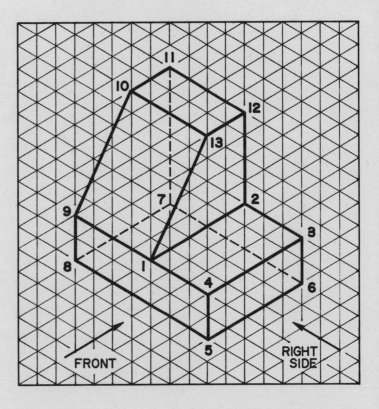

	FRONT VIEW					
Points	View Coordinates			World Coordinates		
	X	Y	Z	X	Y	Z
1						
2						
3						
4						
5						
6						
7						
8						
9						
10						
11						
12						
13						

	RIGHT SIDE VIEW					
Points	View Coordinates			World Coordinates		
	X	Y	Z	X	Y	Z
1						
2						
3						
4						
5						
6						
7						
8						
9						
10						
11						
12						
13						

DRAWN BY   FILE NO.   DRAWING

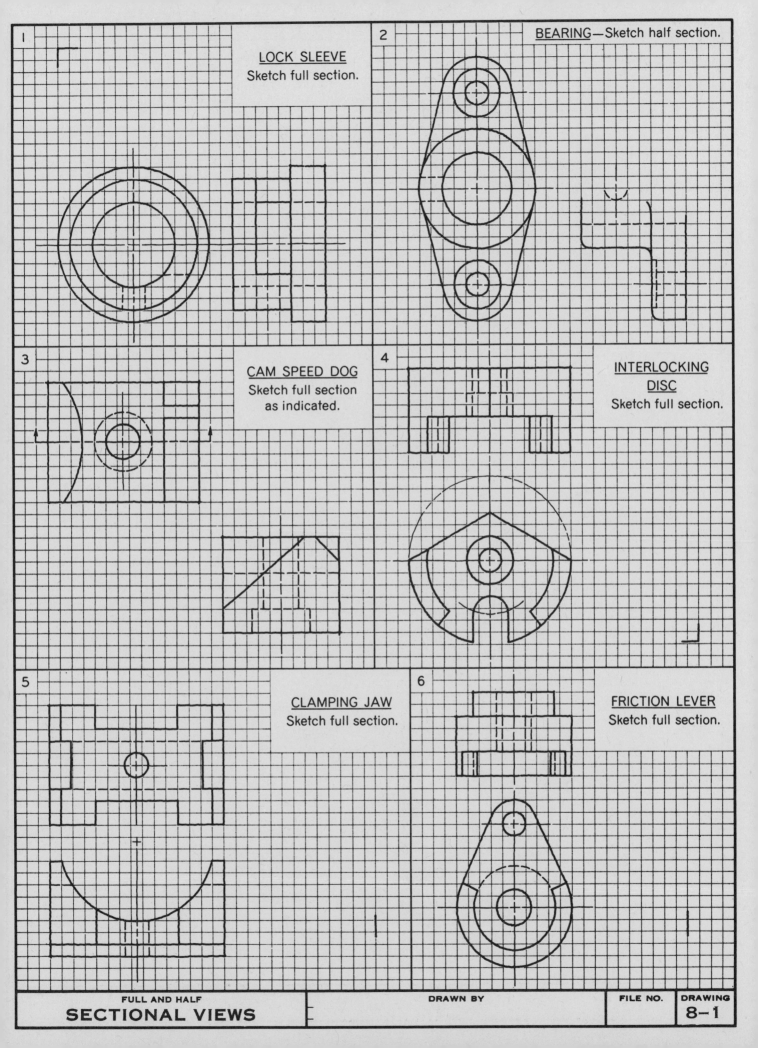

1 LOCK SLEEVE
Sketch full section.

2 BEARING—Sketch half section.

3 CAM SPEED DOG
Sketch full section
as indicated.

4 INTERLOCKING
DISC
Sketch full section.

5 CLAMPING JAW
Sketch full section.

6 FRICTION LEVER
Sketch full section.

FULL AND HALF
SECTIONAL VIEWS

DRAWN BY

FILE NO.

DRAWING
8—1

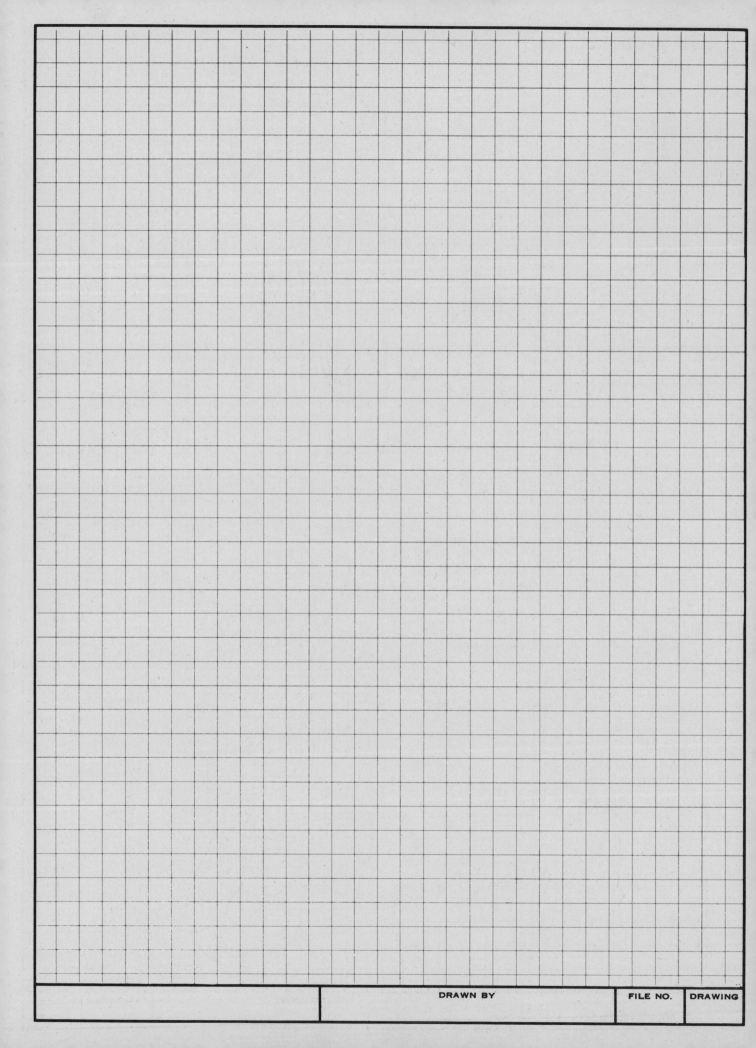

DRAWN BY

FILE NO.

DRAWING

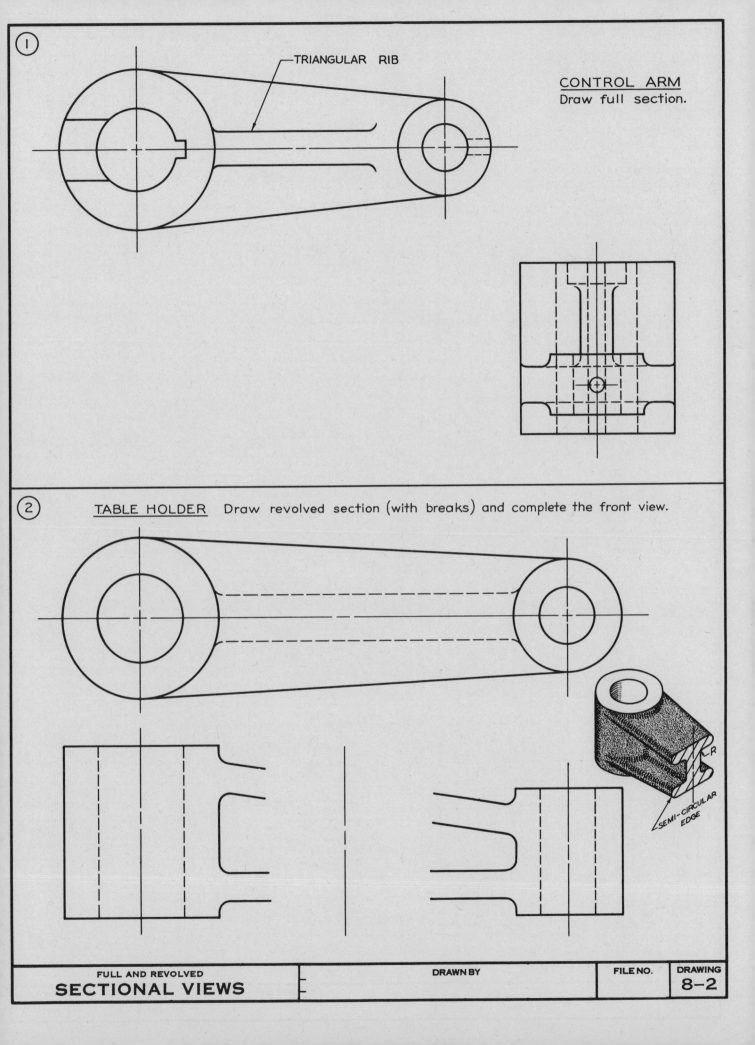

① TRIANGULAR RIB

CONTROL ARM
Draw full section.

② TABLE HOLDER   Draw revolved section (with breaks) and complete the front view.

R

SEMI-CIRCULAR EDGE

FULL AND REVOLVED
SECTIONAL VIEWS

DRAWN BY

FILE NO.

DRAWING
8-2

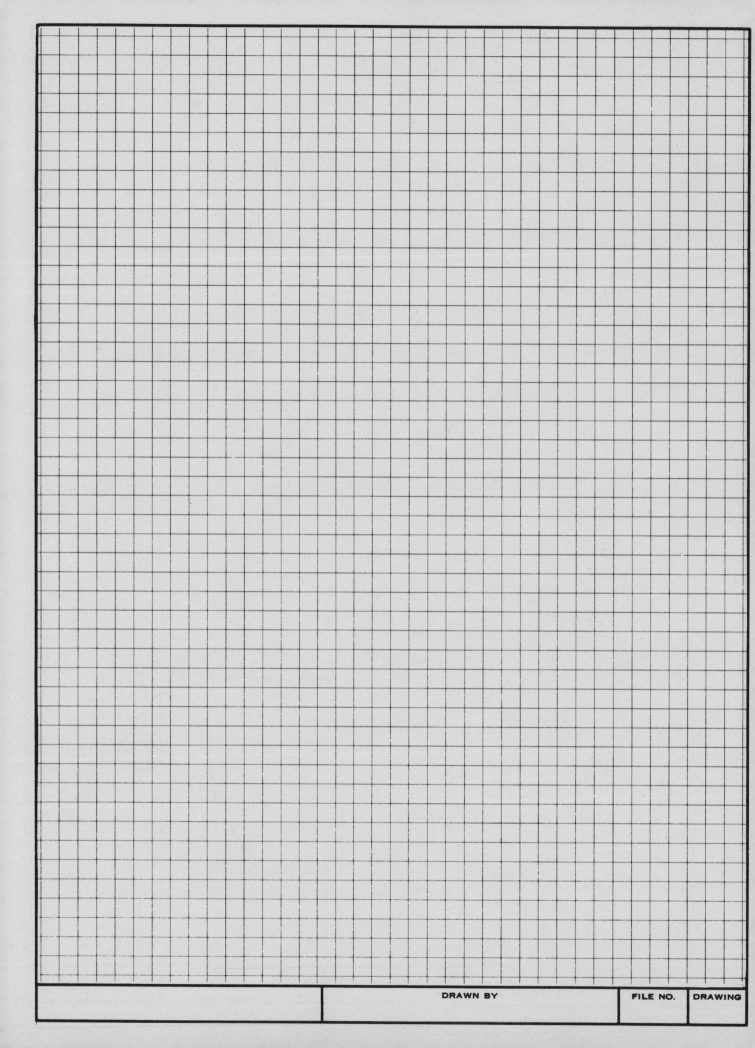

DRAWN BY   FILE NO.   DRAWING

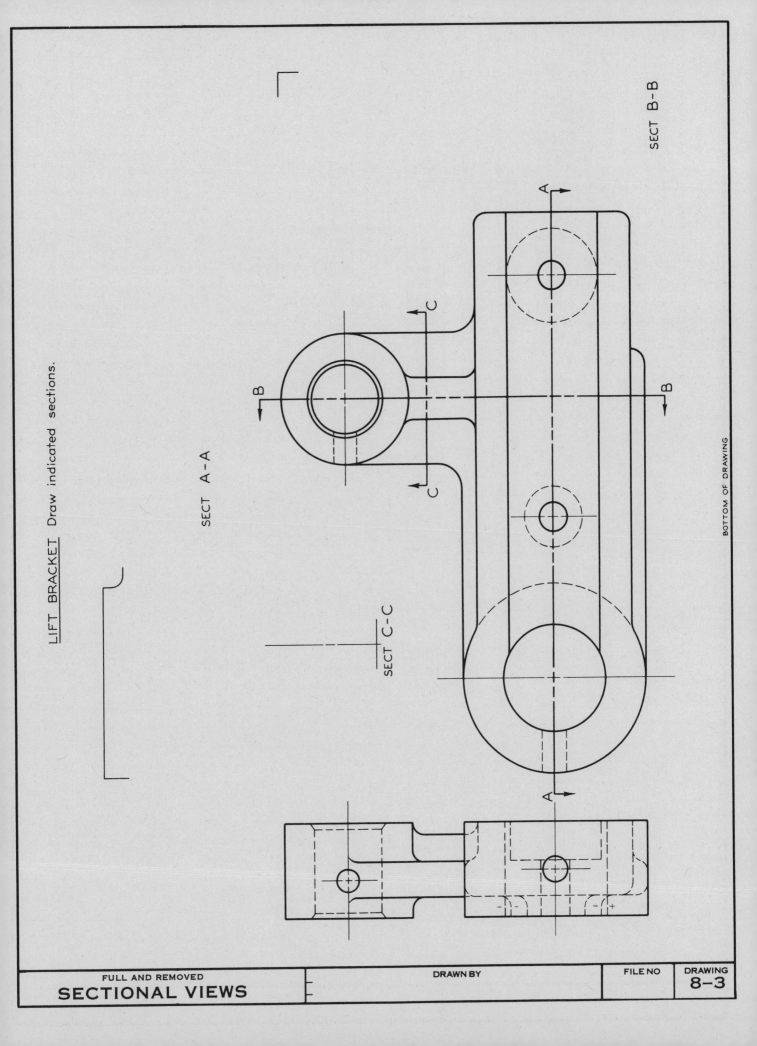

LIFT BRACKET Draw indicated sections.

SECT A-A

SECT C-C

SECT B-B

BOTTOM OF DRAWING

FULL AND REMOVED
SECTIONAL VIEWS

DRAWN BY

FILE NO

DRAWING
8-3

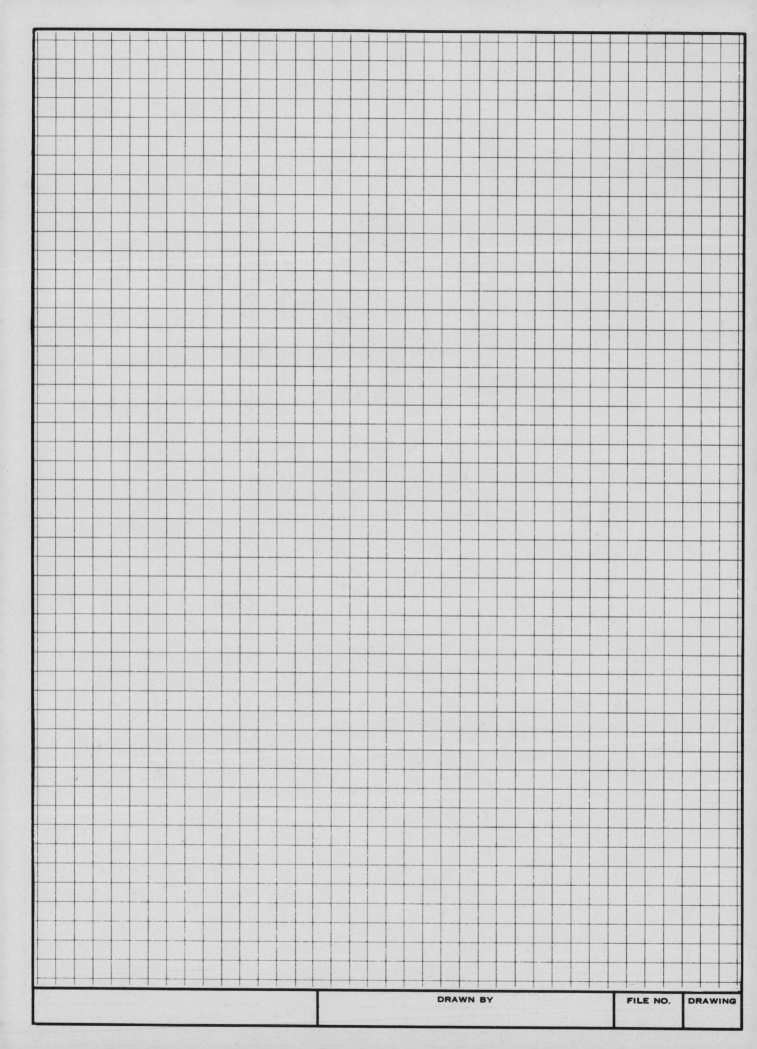

DRAWN BY   FILE NO.   DRAWING

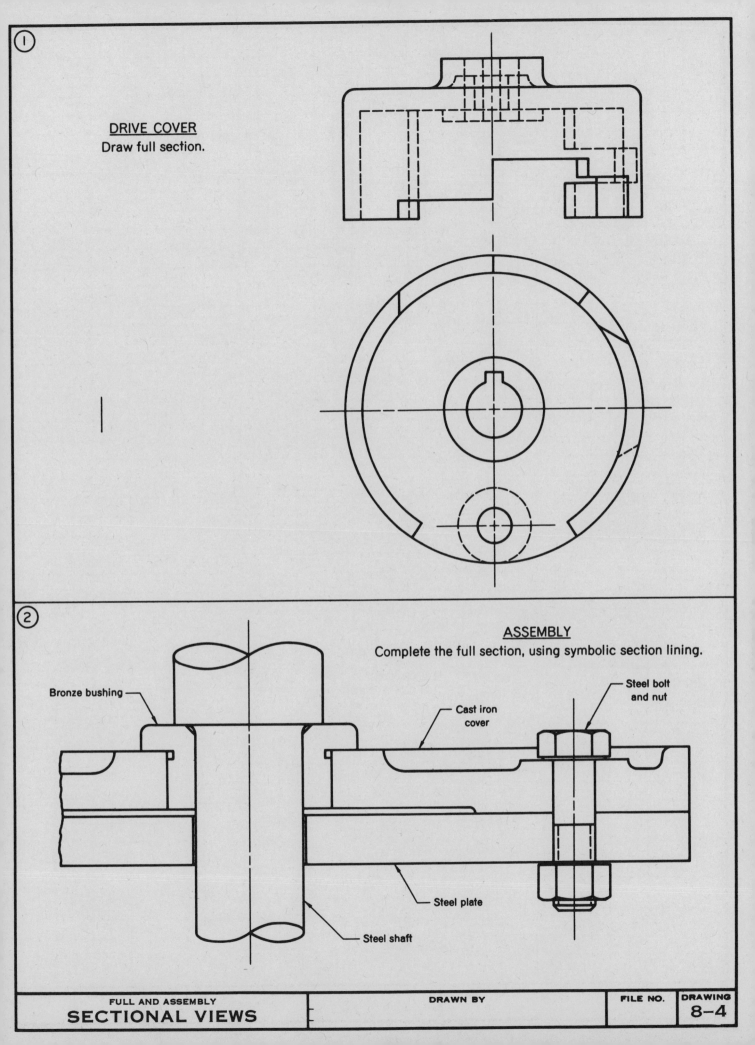

① DRIVE COVER
Draw full section.

② ASSEMBLY
Complete the full section, using symbolic section lining.

Bronze bushing

Cast iron cover

Steel bolt and nut

Steel plate

Steel shaft

FULL AND ASSEMBLY
SECTIONAL VIEWS

DRAWN BY

FILE NO.

DRAWING
8-4

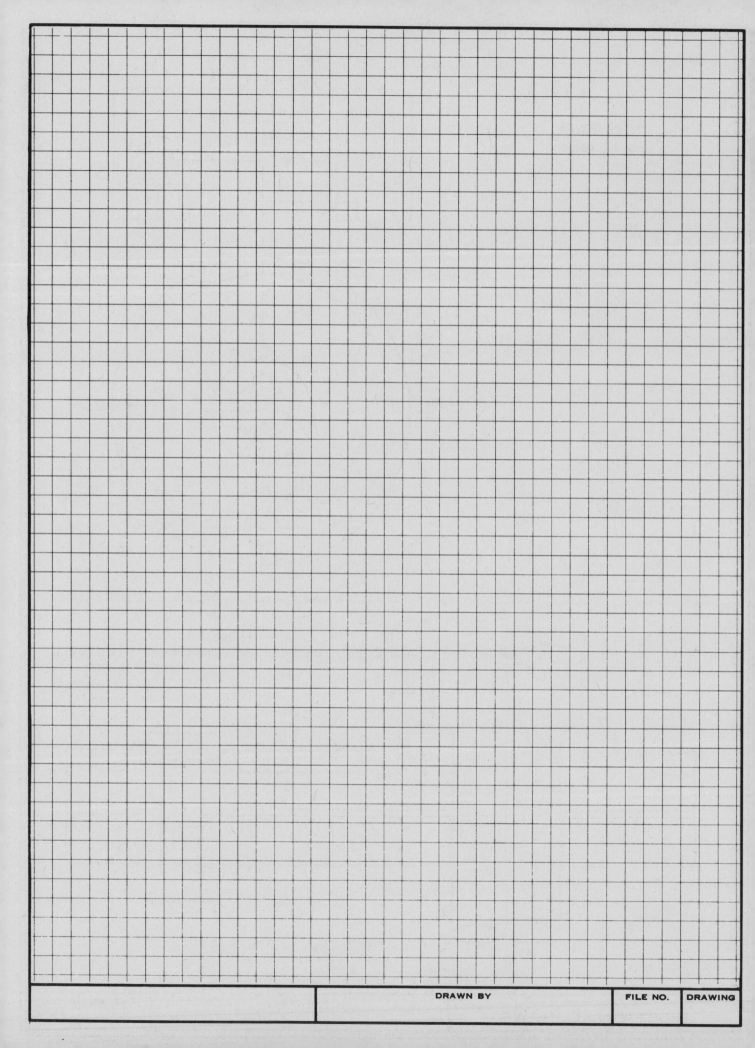

DRAWN BY

FILE NO.

DRAWING

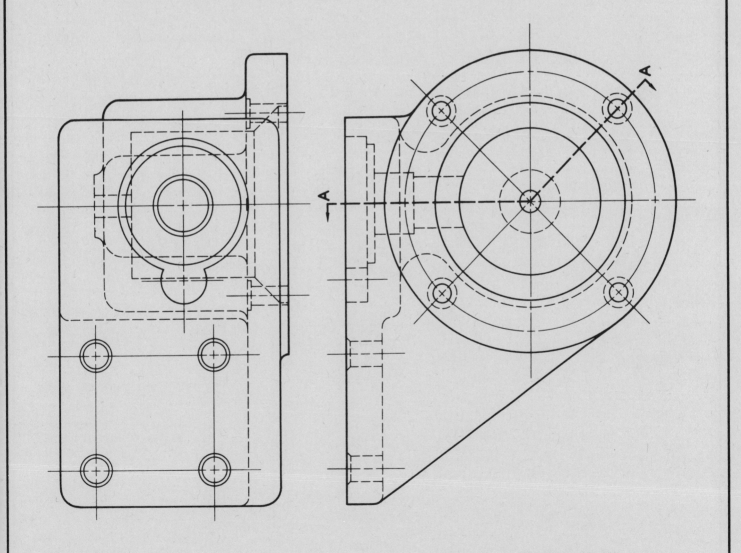

TIMER HOUSING
Draw Section A-A.

A

A

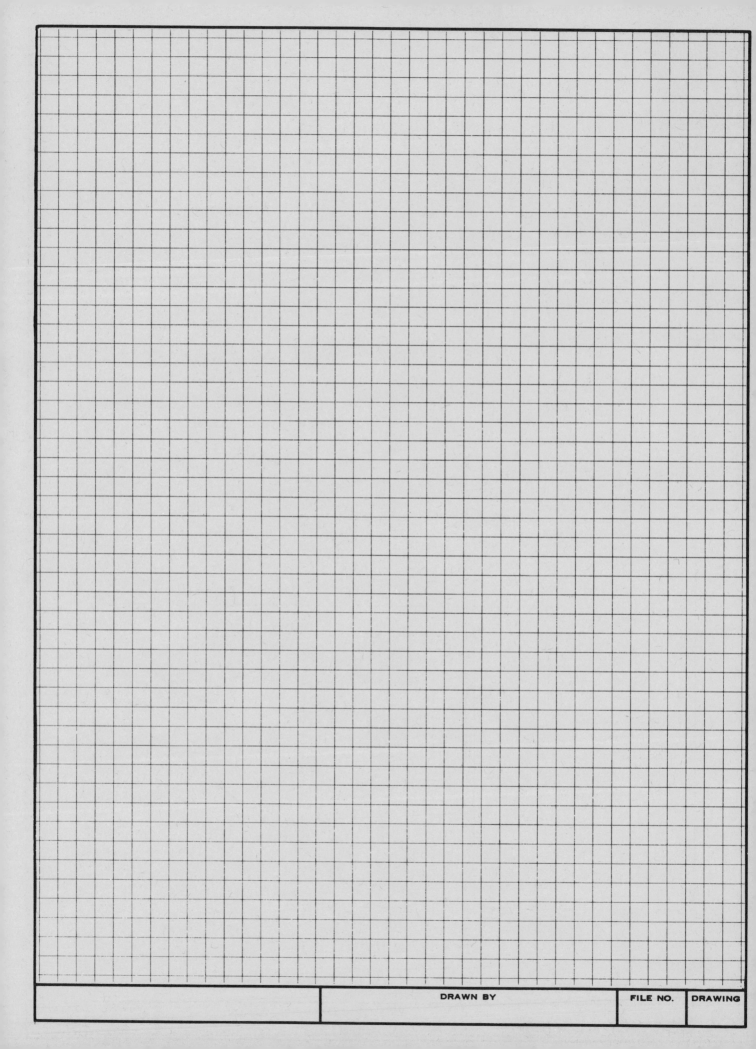

DRAWN BY
FILE NO.
DRAWING

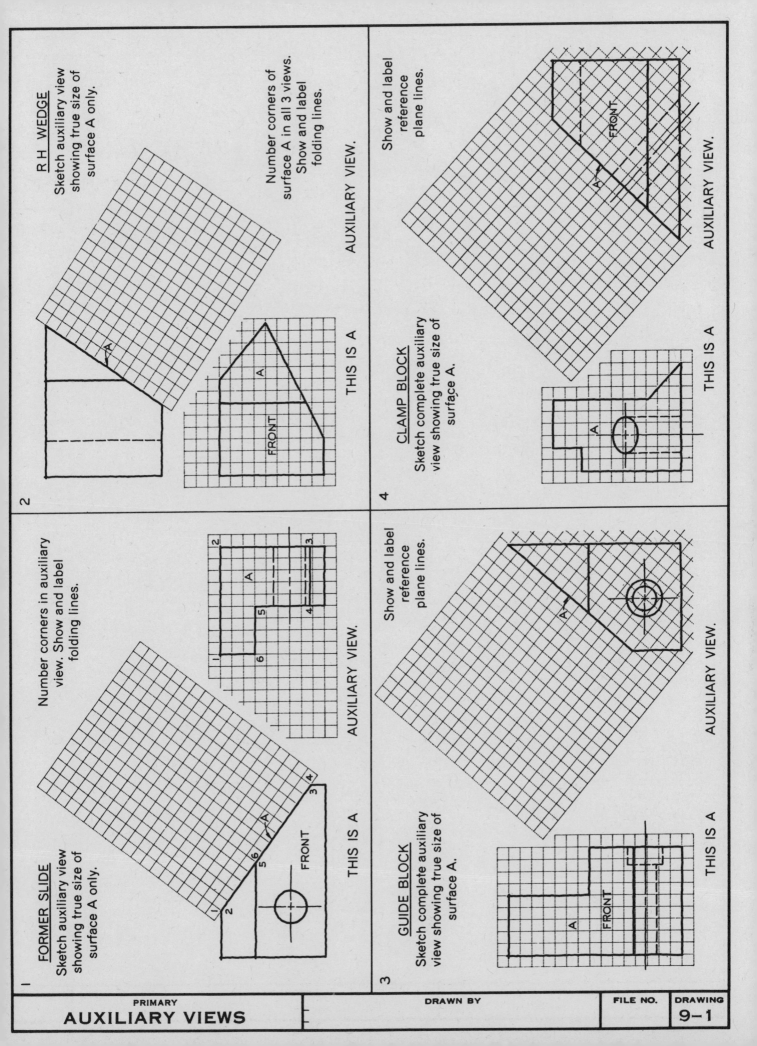

1 FORMER SLIDE
Sketch auxiliary view showing true size of surface A only.

FRONT

THIS IS A

Number corners in auxiliary view. Show and label folding lines.

A

AUXILIARY VIEW.

2 R.H. WEDGE
Sketch auxiliary view showing true size of surface A only.

FRONT

A

THIS IS A

Number corners of surface A in all 3 views. Show and label folding lines.

AUXILIARY VIEW.

3 GUIDE BLOCK
Sketch complete auxiliary view showing true size of surface A.

A

FRONT

THIS IS A

Show and label reference plane lines.

A

AUXILIARY VIEW.

4 CLAMP BLOCK
Sketch complete auxiliary view showing true size of surface A.

A

THIS IS A

Show and label reference plane lines.

FRONT

A

AUXILIARY VIEW.

PRIMARY
AUXILIARY VIEWS

DRAWN BY

FILE NO.

DRAWING
9–1

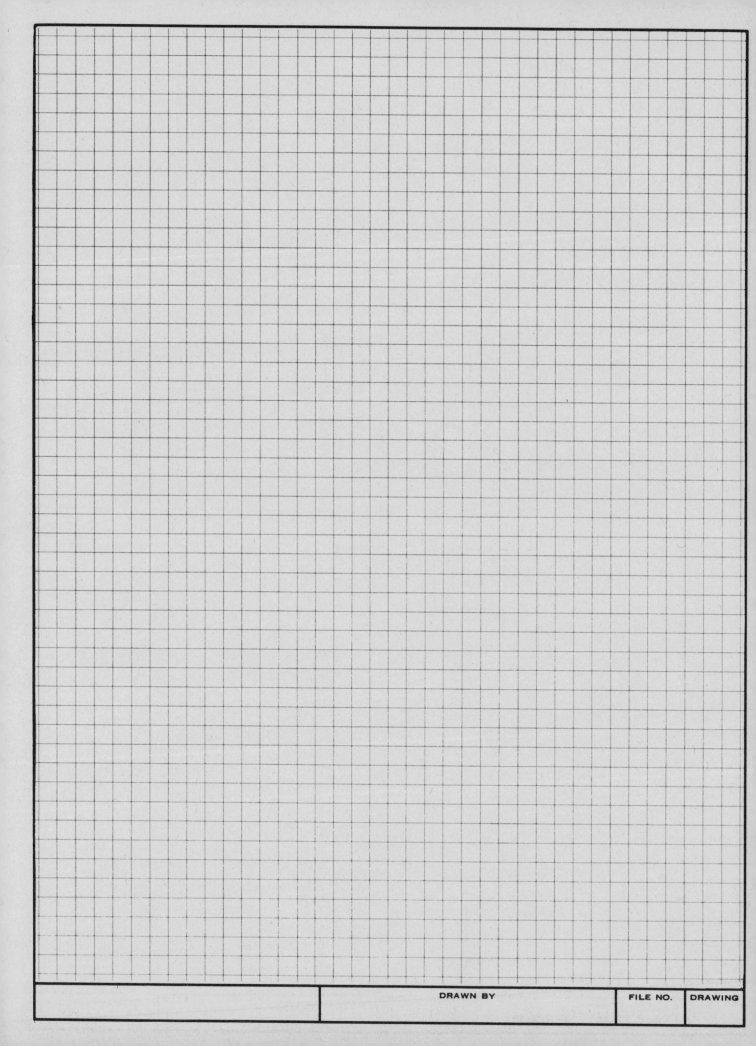

DRAWN BY

FILE NO.   DRAWING

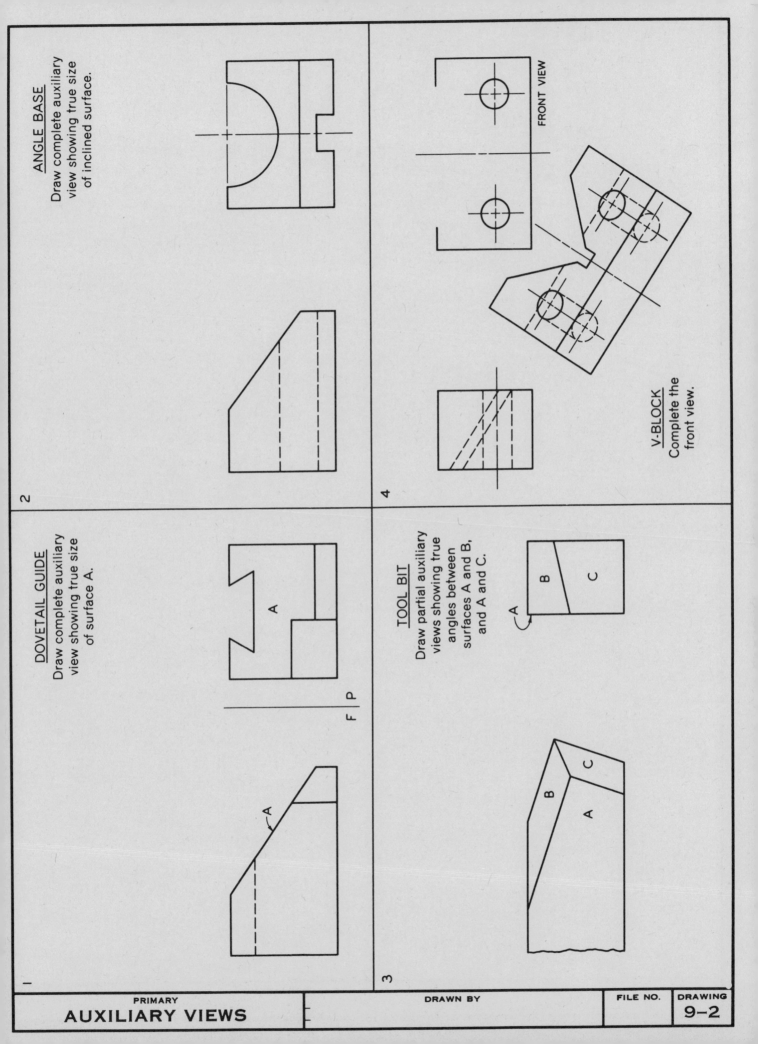

2

ANGLE BASE
Draw complete auxiliary view showing true size of inclined surface.

FRONT VIEW

V-BLOCK
Complete the front view.

4

1

DOVETAIL GUIDE
Draw complete auxiliary view showing true size of surface A.

A

F/P

A

TOOL BIT
Draw partial auxiliary views showing true angles between surfaces A and B, and A and C.

A

B

C

B

A

C

3

PRIMARY
AUXILIARY VIEWS

DRAWN BY

FILE NO.

DRAWING
9-2

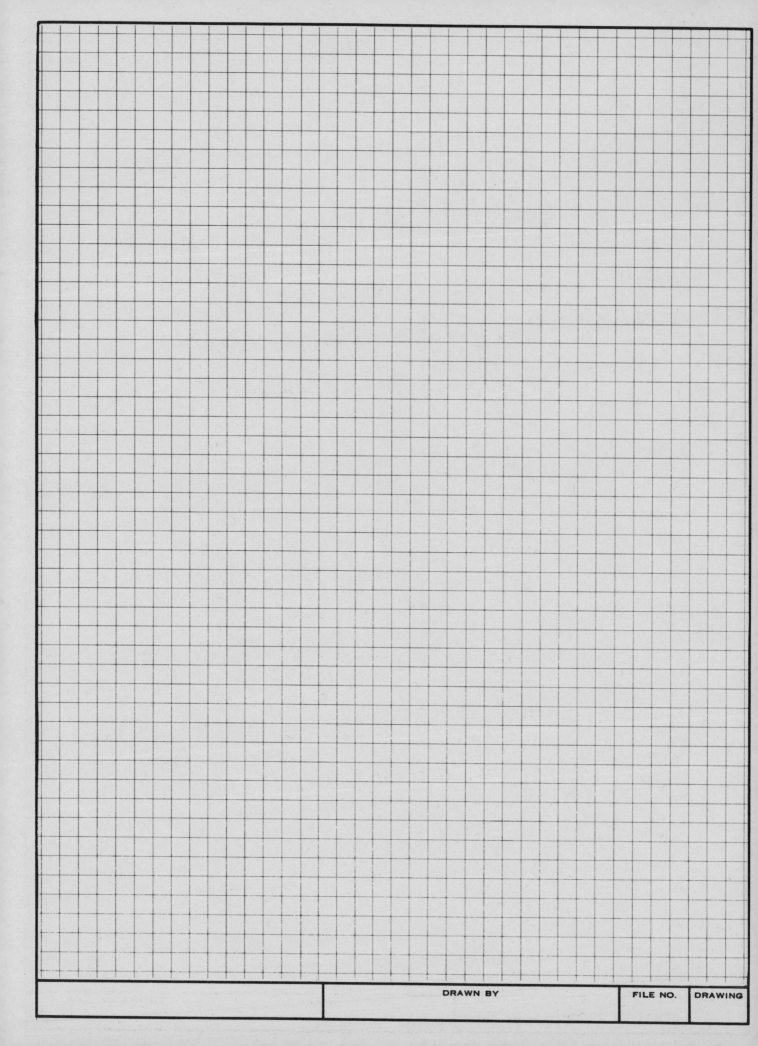

DRAWN BY

FILE NO.

DRAWING

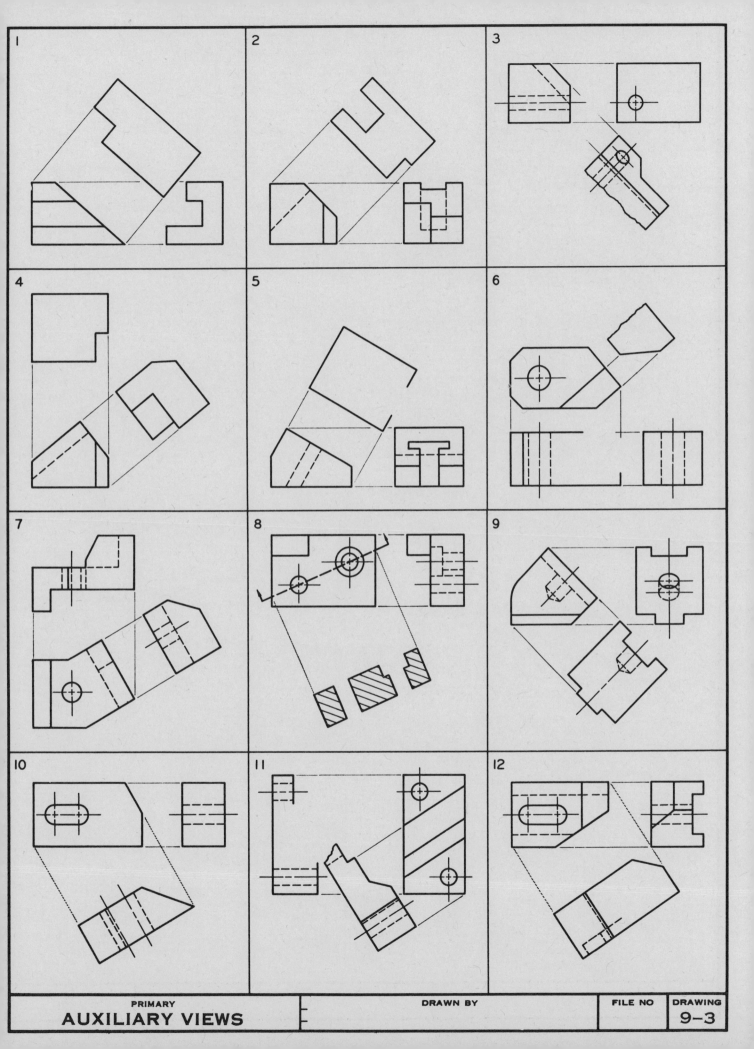

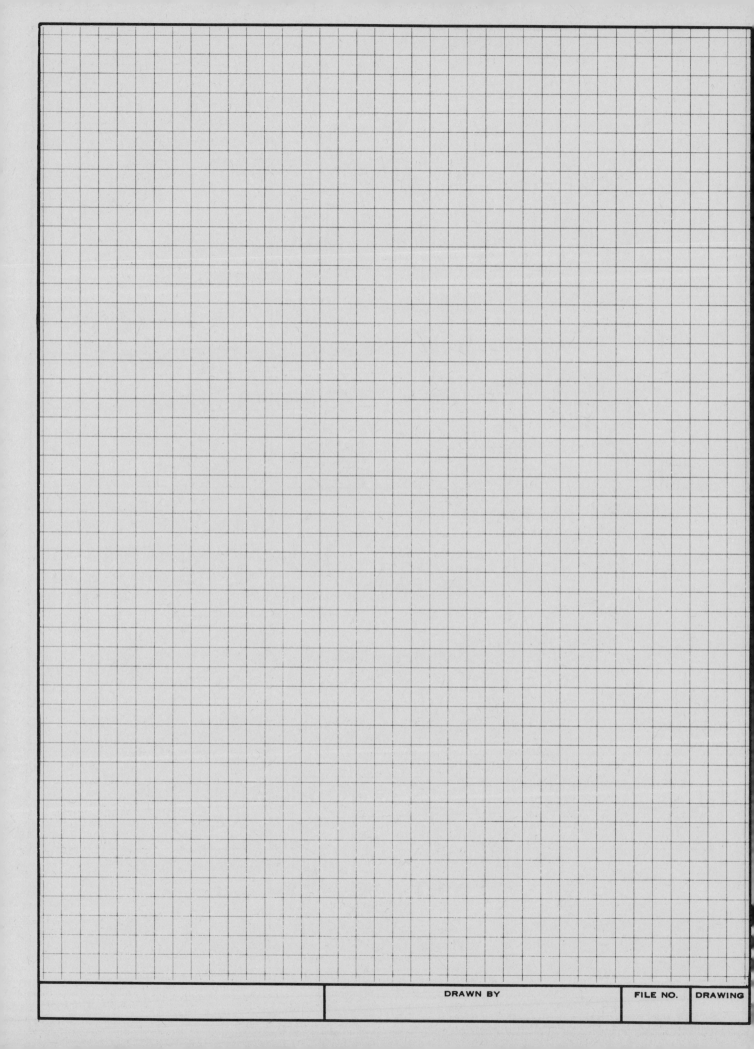

1

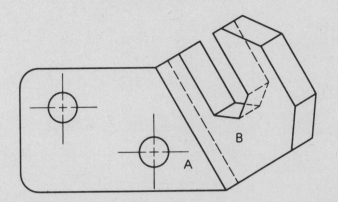

ANGLE BRACKET

Draw complete primary auxiliary view showing 135° angle between surfaces A and B; then draw partial secondary auxiliary view showing true size of member B. Thickness of members-10mm.

2

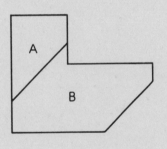

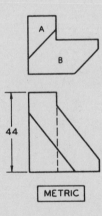

44

METRIC

STOP BLOCK

Draw complete primary auxiliary view showing angle between surfaces A and B; then draw complete secondary auxiliary view showing true size of surface B.

| SECONDARY
AUXILIARY VIEWS | | DRAWN BY | FILE NO. | DRAWING
9-4 |

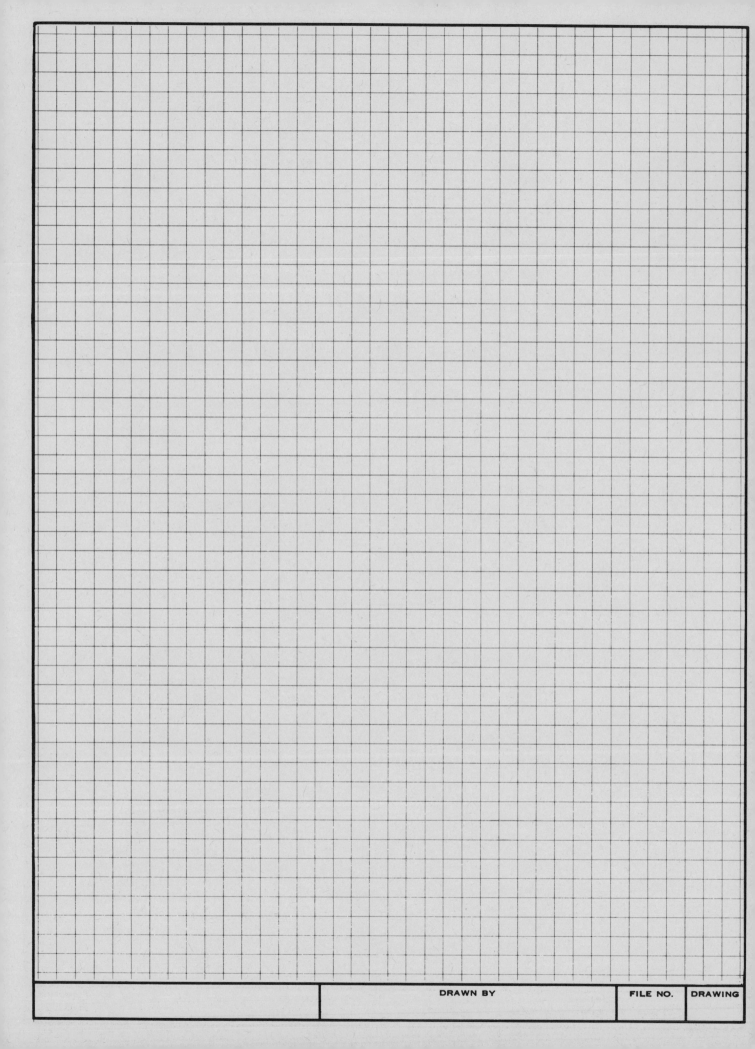

DRAWN BY

FILE NO.

DRAWING

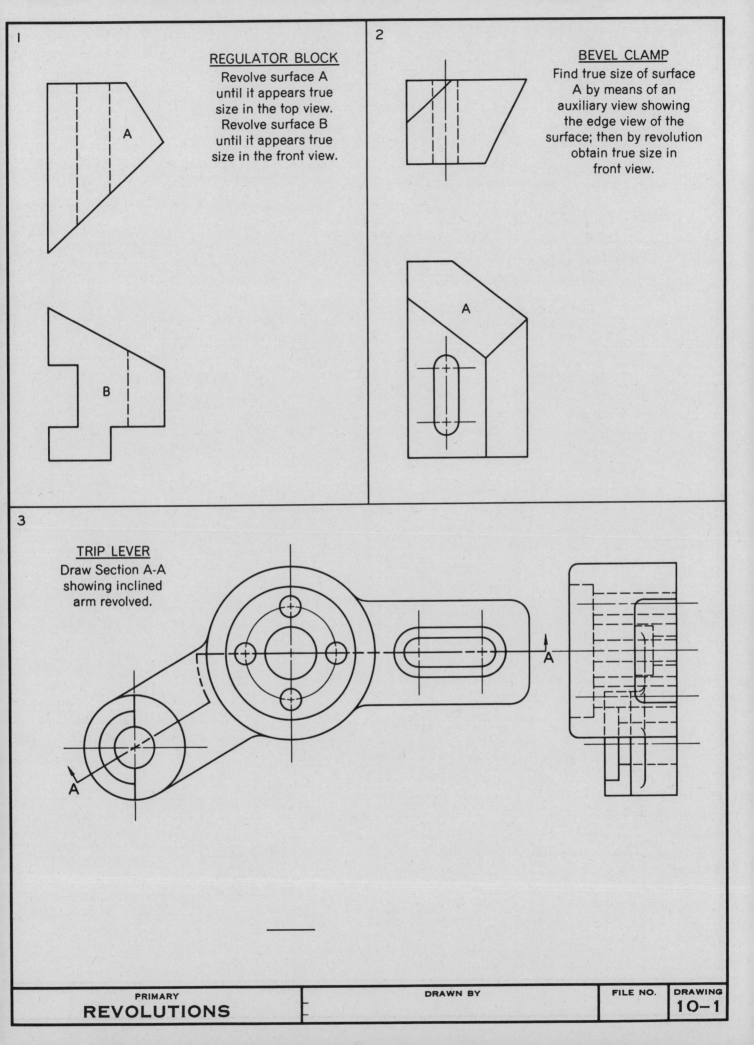

1

<u>REGULATOR BLOCK</u>
Revolve surface A
until it appears true
size in the top view.
Revolve surface B
until it appears true
size in the front view.

2

<u>BEVEL CLAMP</u>
Find true size of surface
A by means of an
auxiliary view showing
the edge view of the
surface; then by revolution
obtain true size in
front view.

3

<u>TRIP LEVER</u>
Draw Section A-A
showing inclined
arm revolved.

| PRIMARY **REVOLUTIONS** | | DRAWN BY | FILE NO. | DRAWING **10-1** |

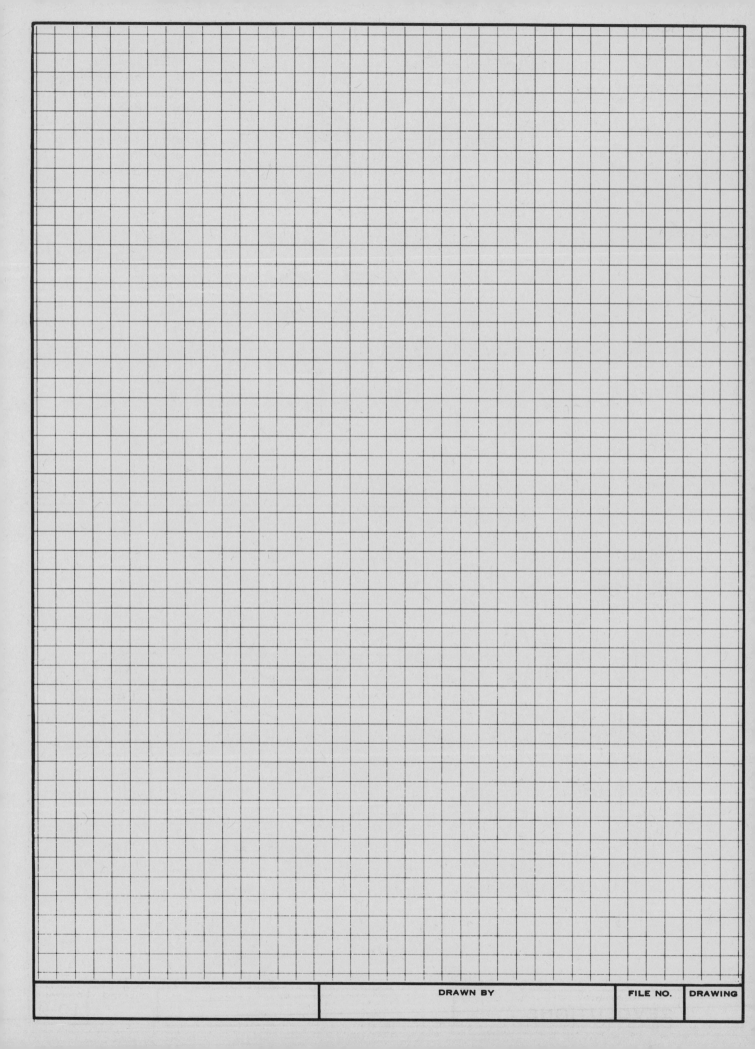

DRAWN BY

FILE NO.

DRAWING

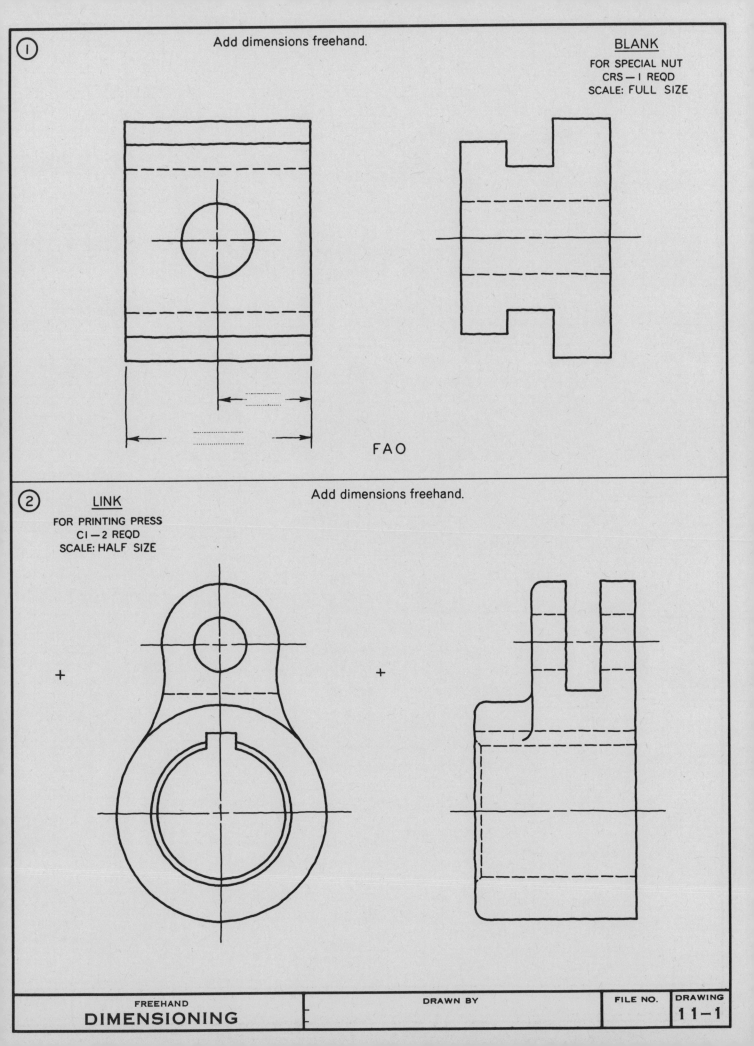

① Add dimensions freehand.

BLANK
FOR SPECIAL NUT
CRS — I REQD
SCALE: FULL SIZE

FAO

② Add dimensions freehand.

LINK
FOR PRINTING PRESS
CI — 2 REQD
SCALE: HALF SIZE

FREEHAND
DIMENSIONING

DRAWN BY

FILE NO.

DRAWING
11-1

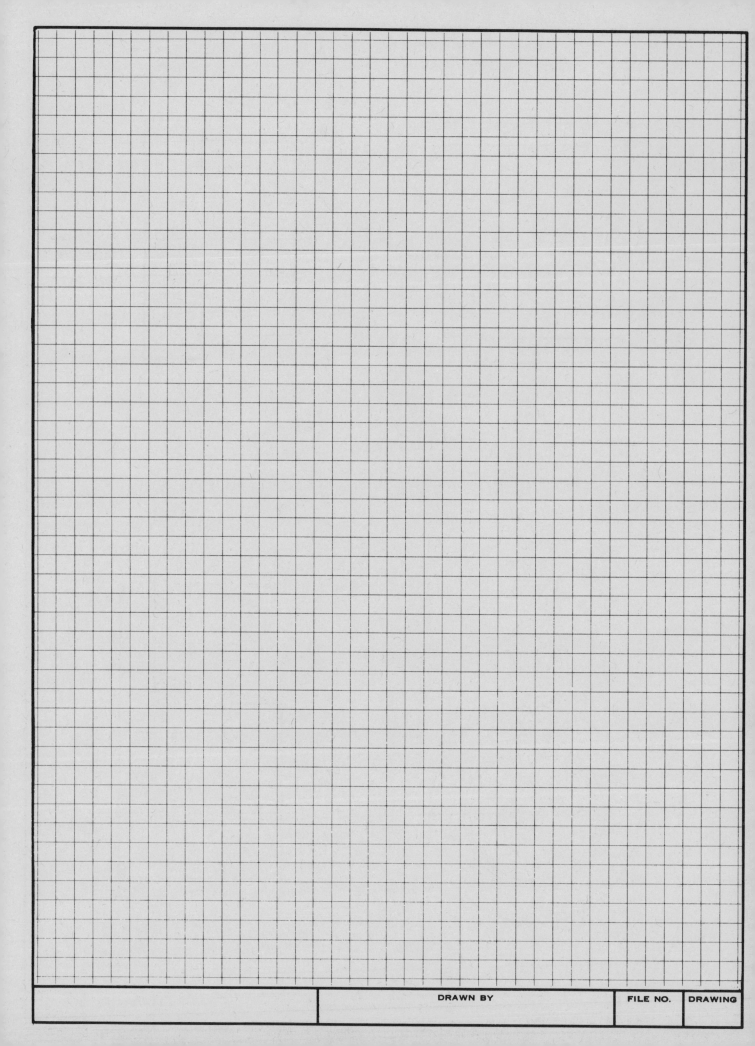

DRAWN BY

FILE NO.

DRAWING

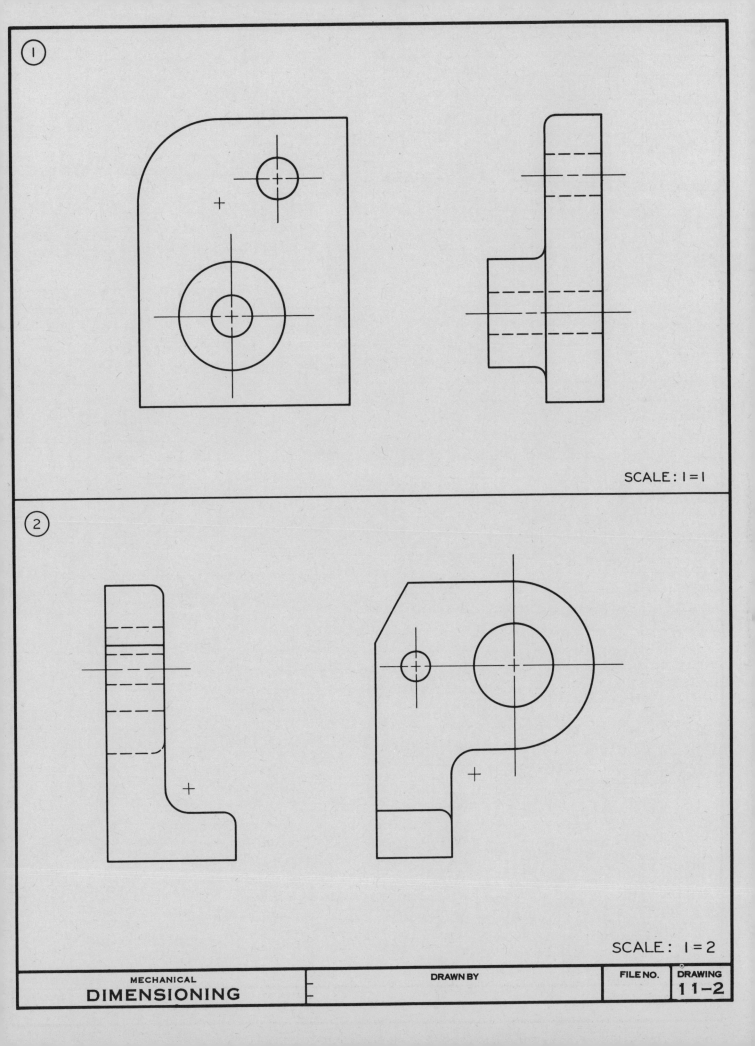

① SCALE: 1=1

② SCALE: 1=2

MECHANICAL
DIMENSIONING

DRAWN BY

FILE NO.

DRAWING
11-2

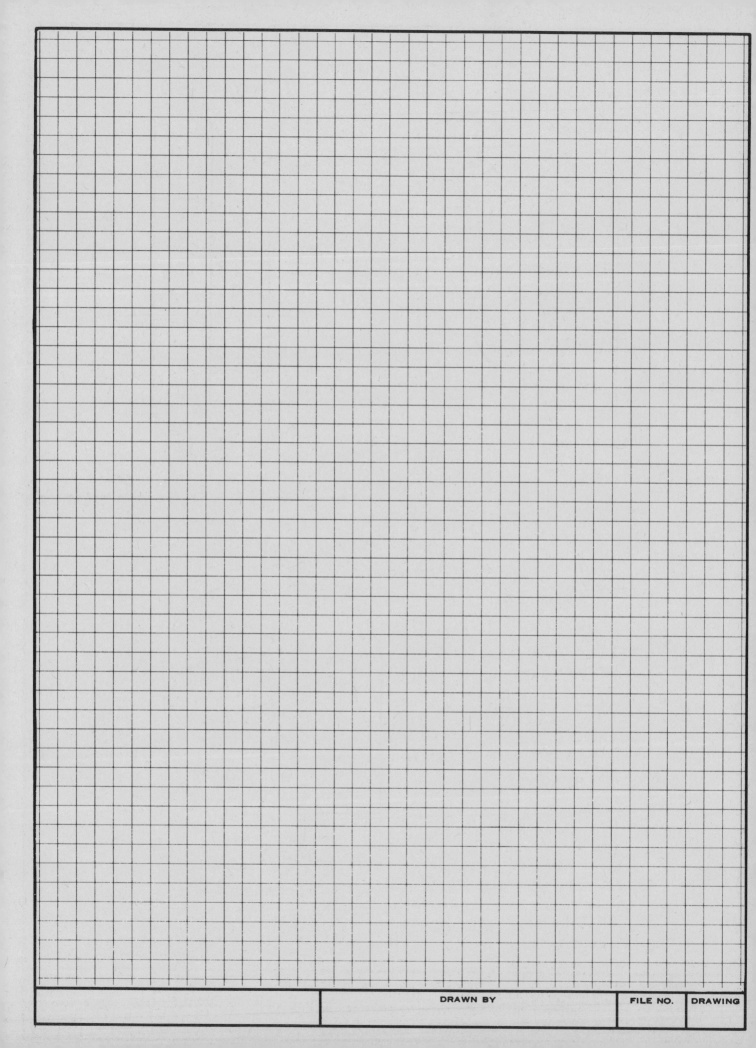

DRAWN BY

FILE NO.

DRAWING

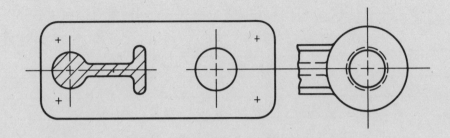

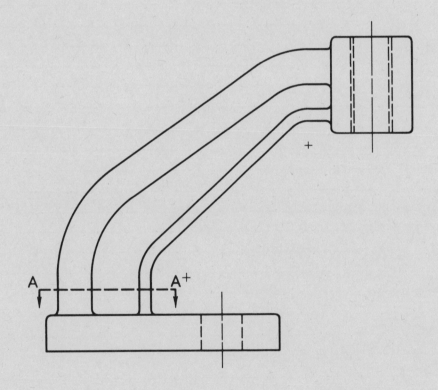

A

A

DRAWN BY   FILE NO.   DRAWING

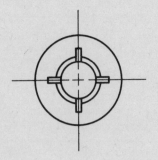

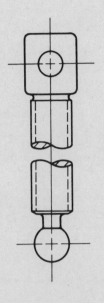

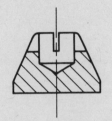

③ PAD   CRS - 1 REQD

② CLAMP SCREW
CRS - 1 REQD

⑤ HANDLE CAP
CRS - 2 REQD

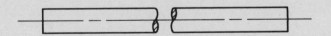

④ HANDLE   CRS - 1 REQD

STANDARD PARTS

⑥ 1 - M10 x 1.5
HEAVY HEX NUT

⑦ 1 - 12 x 22 x 4.5 T-SLOT
WASHER

⑧ 1 - M10 x 1.5 - 50 LONG
T-SLOT BOLT

UNSPECIFIED FILLETS & ROUNDS 1.5R   FULL SIZE

| MATING PARTS | DRAWN BY | FILE NO. | DRAWING |
| **DIMENSIONING** | | | 11-4 |

DRAWN BY   FILE NO.   DRAWING

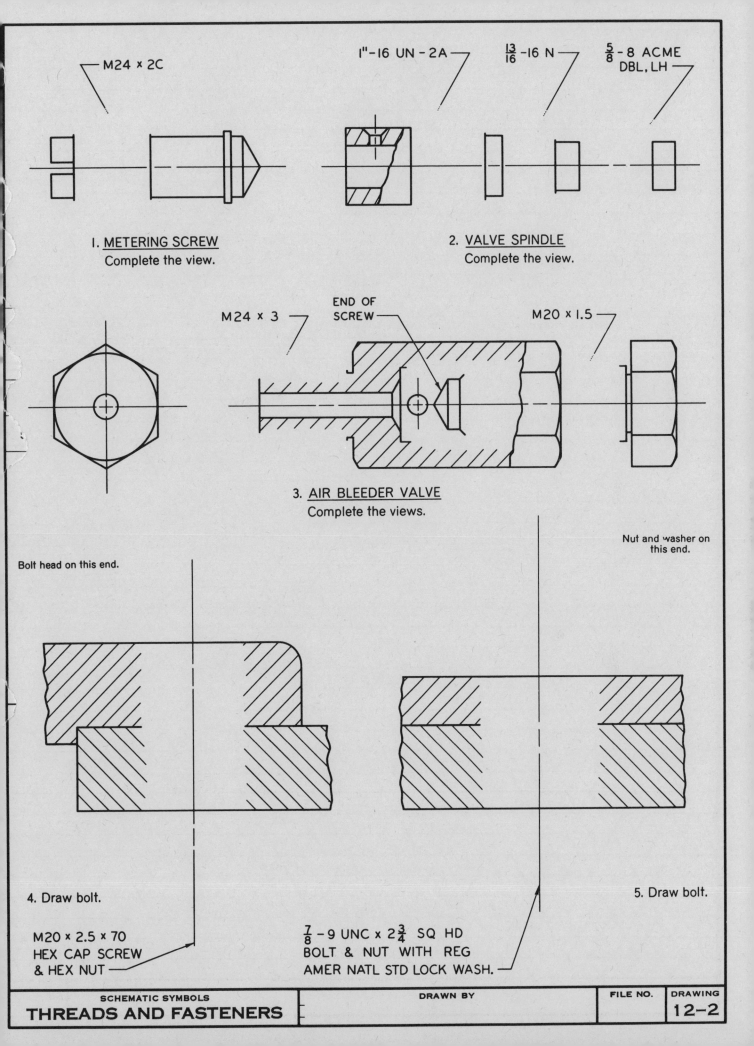

M24 × 2C

1. METERING SCREW
Complete the view.

1"-16 UN - 2A   13/16 -16 N   5/8 - 8 ACME DBL, LH

2. VALVE SPINDLE
Complete the view.

M24 × 3   END OF SCREW   M20 × 1.5

3. AIR BLEEDER VALVE
Complete the views.

Nut and washer on this end.

Bolt head on this end.

4. Draw bolt.

M20 × 2.5 × 70
HEX CAP SCREW
& HEX NUT

5. Draw bolt.

7/8 - 9 UNC × 2 3/4 SQ HD
BOLT & NUT WITH REG
AMER NATL STD LOCK WASH.

DRAWN BY · FILE NO. · DRAWING

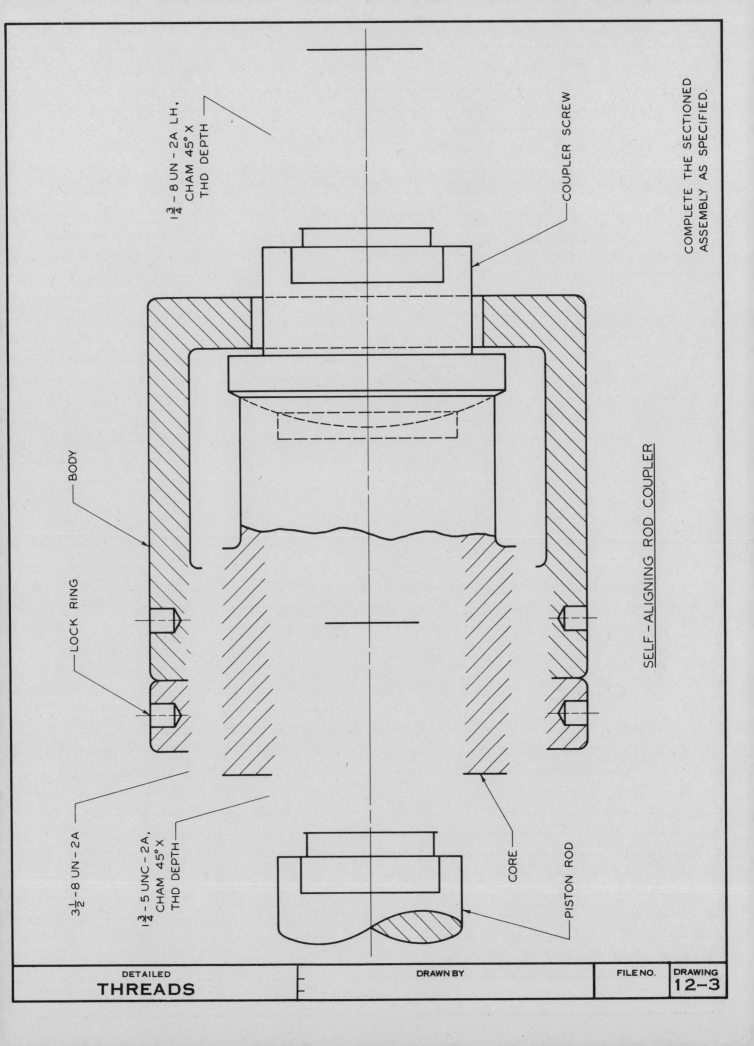

$1\frac{3}{4} - 8$ UN - 2A LH,
CHAM 45° X
THD DEPTH

COUPLER SCREW

BODY

LOCK RING

SELF-ALIGNING ROD COUPLER

COMPLETE THE SECTIONED
ASSEMBLY AS SPECIFIED.

$3\frac{1}{2} - 8$ UN - 2A

$1\frac{3}{4} - 5$ UNC - 2A,
CHAM 45° X
THD DEPTH

CORE

PISTON ROD

| DETAILED | DRAWN BY | FILE NO. | DRAWING |
| THREADS | | | 12-3 |

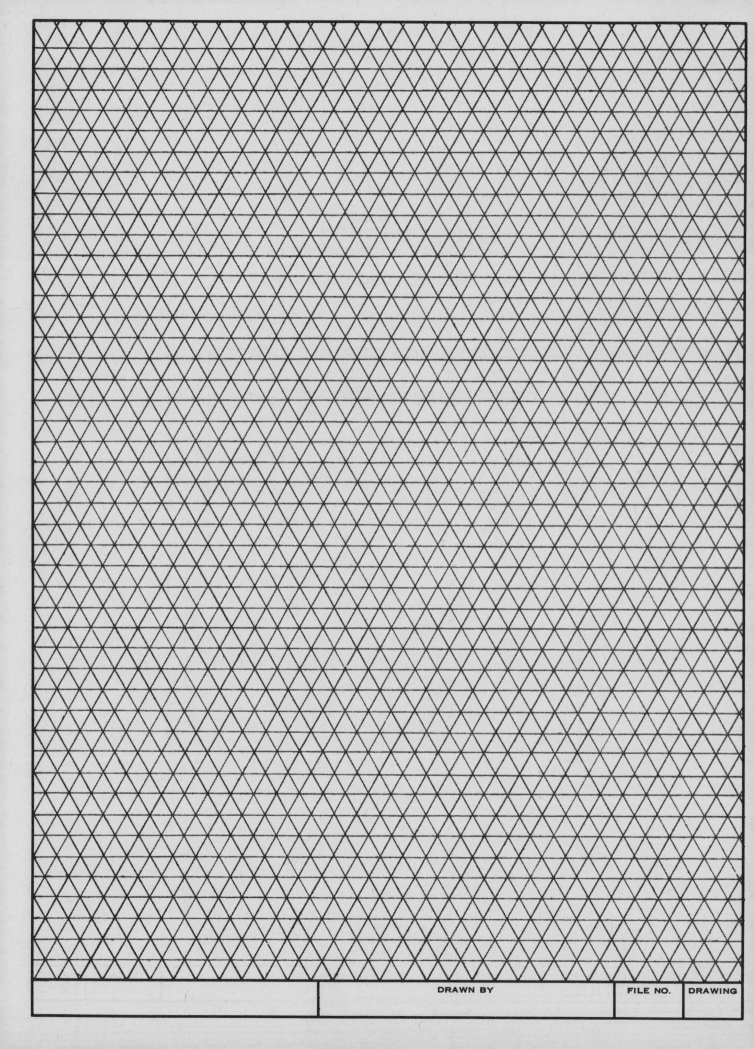

DRAWN BY

FILE NO.

DRAWING

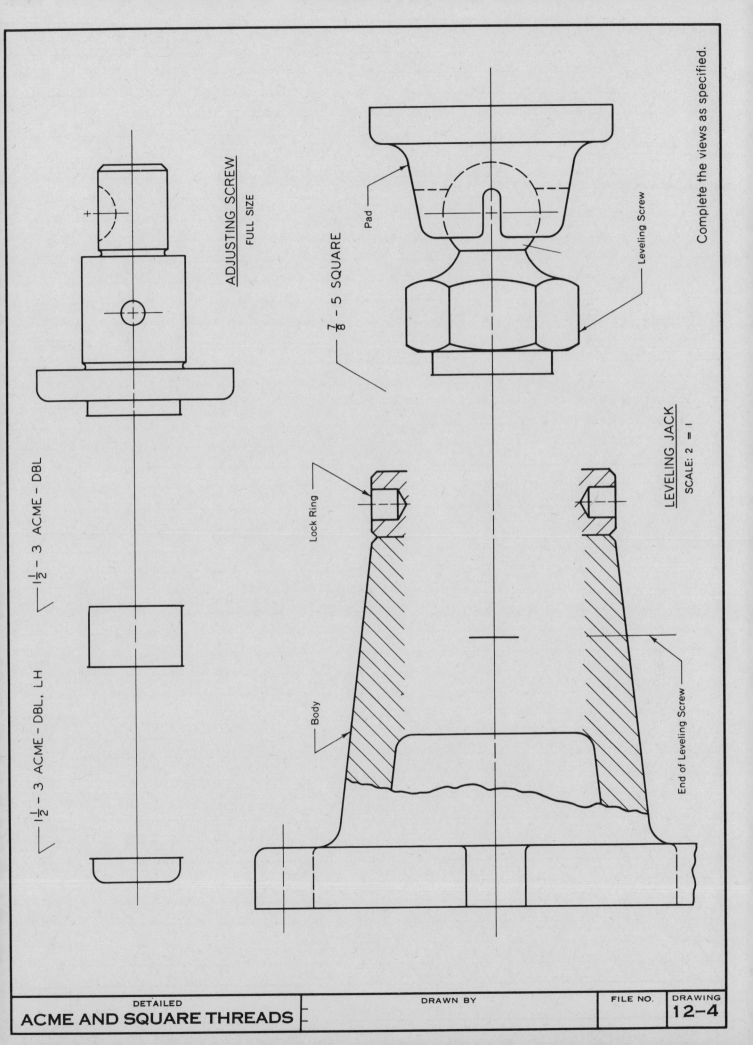

ADJUSTING SCREW

FULL SIZE

$\frac{7}{8}$ - 5 SQUARE

Pad

Leveling Screw

Complete the views as specified.

$1\frac{1}{2}$ - 3 ACME - DBL

$1\frac{1}{2}$ - 3 ACME - DBL, LH

LEVELING JACK

SCALE: 2 = 1

Lock Ring

Body

End of Leveling Screw

DETAILED

ACME AND SQUARE THREADS

DRAWN BY

FILE NO.

DRAWING

12-4

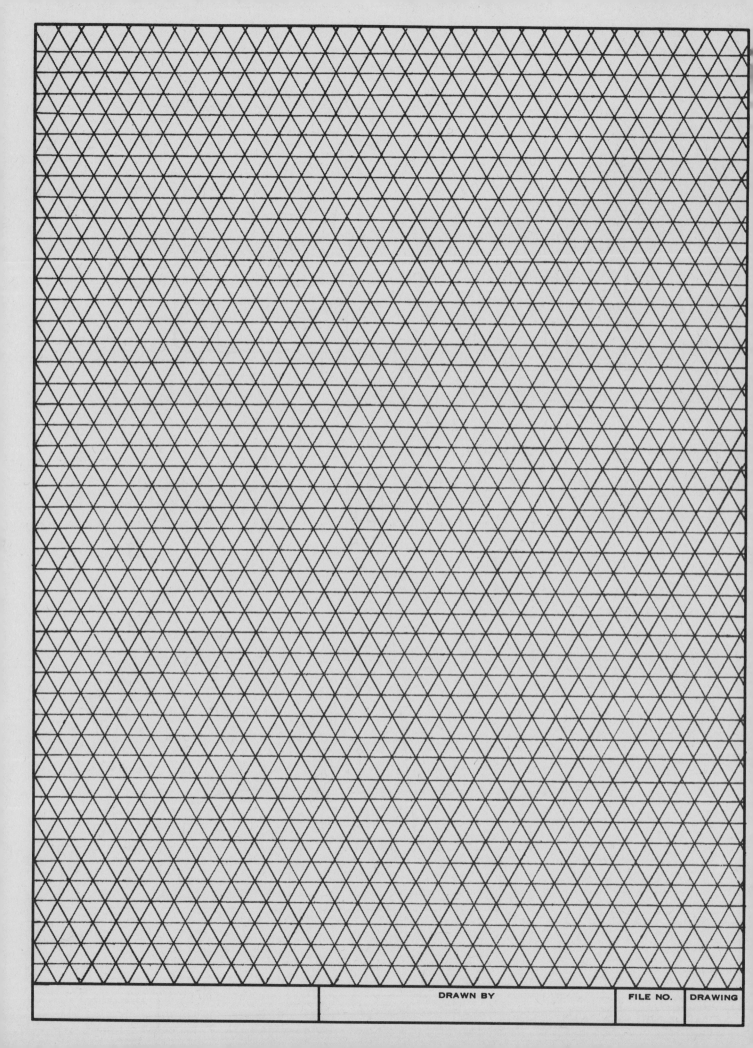

DRAWN BY   FILE NO.   DRAWING

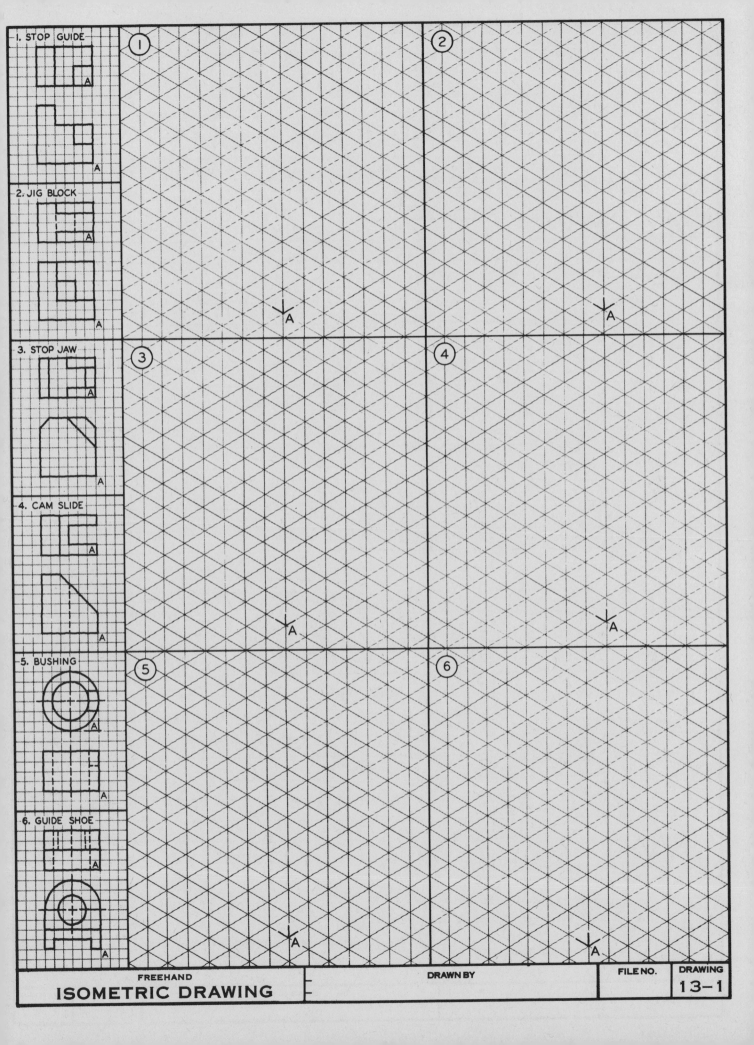

1. STOP GUIDE

A

A

2. JIG BLOCK

A

A

3. STOP JAW

A

A

4. CAM SLIDE

A

A

5. BUSHING

A

A

6. GUIDE SHOE

A

①   ②

A   A

③   ④

A   A

⑤   ⑥

A   A

FREEHAND
ISOMETRIC DRAWING

DRAWN BY

FILE NO.

DRAWING
13-1

DRAWN BY   FILE NO.   DRAWING

ISOMETRIC DRAWING

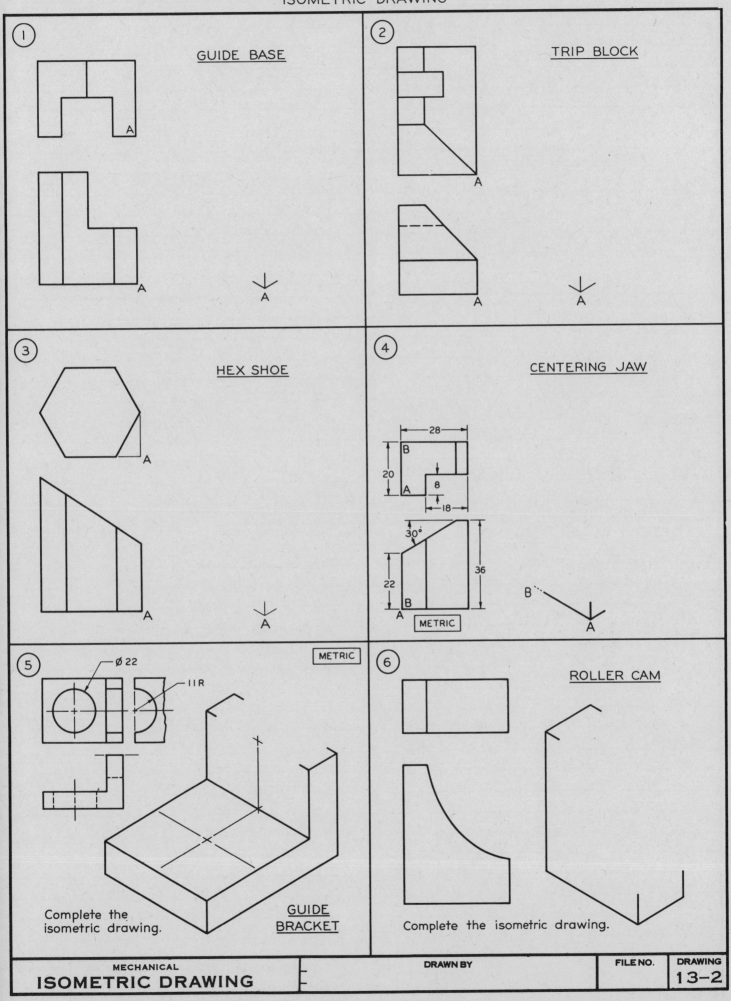

1 GUIDE BASE

A

A

A

2 TRIP BLOCK

A

A

A

3 HEX SHOE

A

A

A

4 CENTERING JAW

28

B

20

A   8

18

30°

22   36

B

A   METRIC

B

A

5 Ø 22

11 R

METRIC

Complete the isometric drawing.

GUIDE BRACKET

6 ROLLER CAM

Complete the isometric drawing.

MECHANICAL
ISOMETRIC DRAWING

DRAWN BY

FILE NO.

DRAWING
13-2

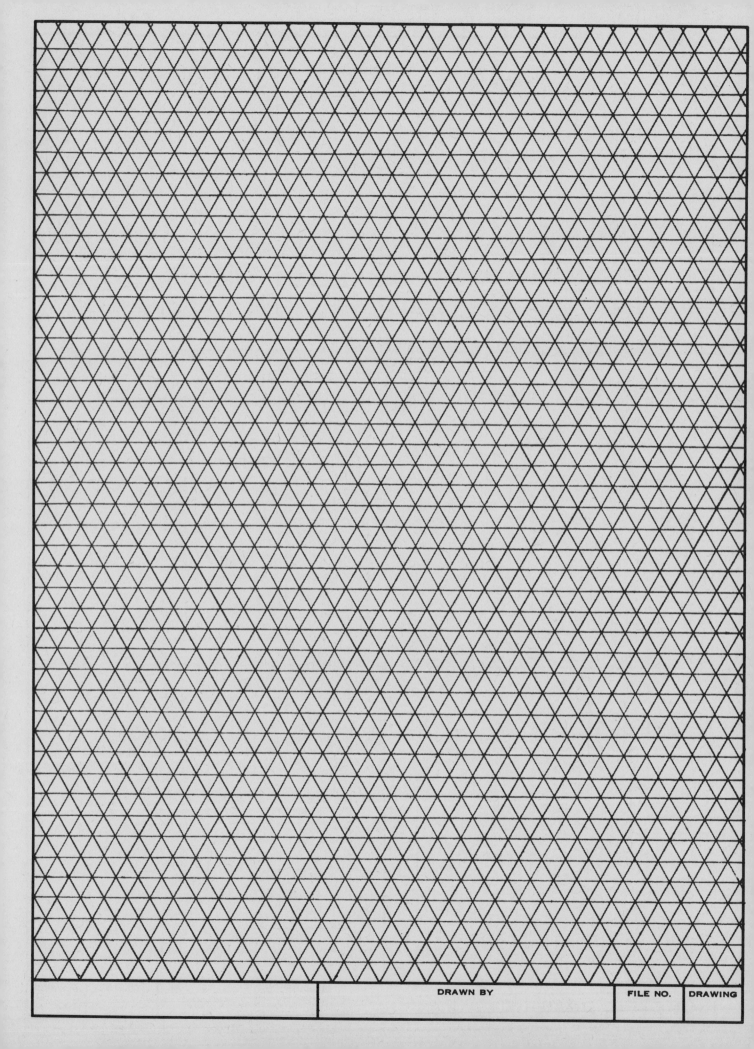

DRAWN BY FILE NO. DRAWING

Make isometric drawing. Omit dimensions.

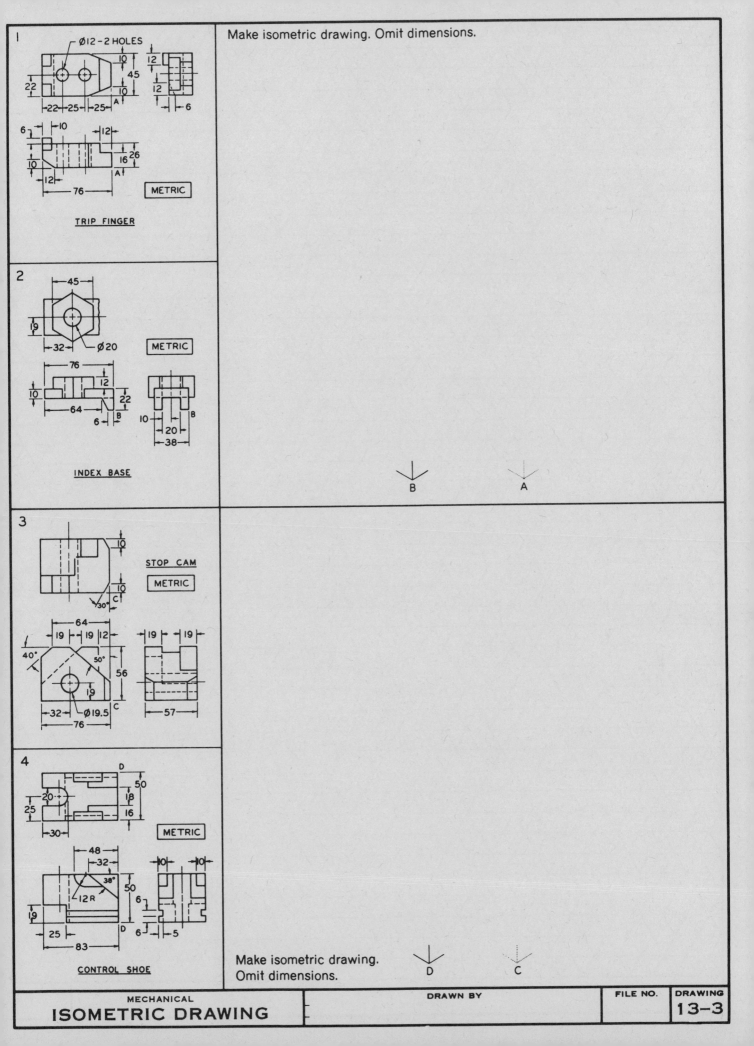

1

Ø12 - 2 HOLES

10

12

22 45 12 12

-22-25-25- A 6

TRIP FINGER

6 10 12

10 26

10 16

12 A

76

METRIC

2

45

19

-32- Ø20

METRIC

76 12

10 22

64 B

6 B

10 20 38

INDEX BASE

3

10

STOP CAM

10

METRIC

30° C

64

19 19 12 19 19

40° 50°

56

19

32 Ø19.5 C 57

76

4

D

50

20 18

25 16

-30- METRIC

48

32

38°

50

12 R 6

19 6

25 D

83 6 5

CONTROL SHOE

B

A

Make isometric drawing.
Omit dimensions.

D

C

MECHANICAL
ISOMETRIC DRAWING

DRAWN BY

FILE NO.

DRAWING
13-3

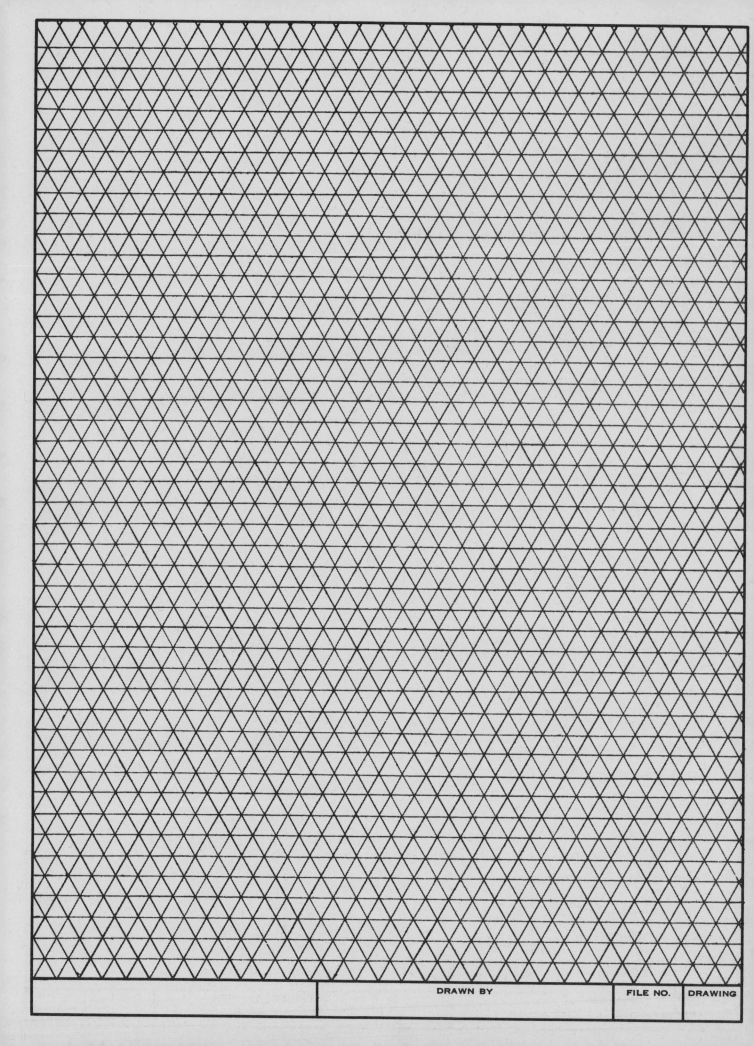

DRAWN BY

FILE NO.

DRAWING

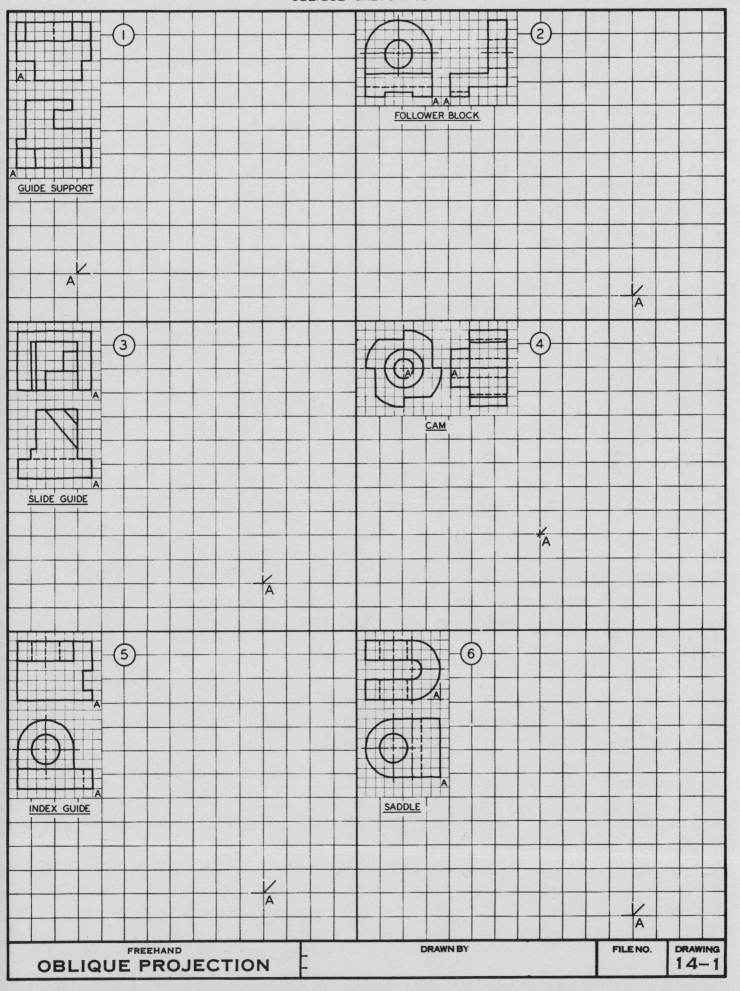

1   A   A   GUIDE SUPPORT

2   A A   FOLLOWER BLOCK   A

3   A   SLIDE GUIDE   A   A

4   A   A   CAM   A

5   A   INDEX GUIDE   A   A

6   A   SADDLE   A   A

DRAWN BY · FILE NO. · DRAWING

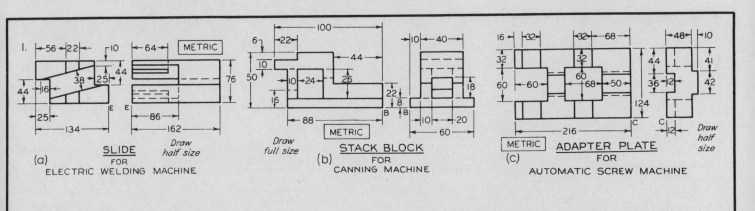

1. SLIDE
FOR
ELECTRIC WELDING MACHINE
(a)
METRIC
Draw
half size

STACK BLOCK
FOR
CANNING MACHINE
(b)
METRIC
Draw
full size

ADAPTER PLATE
FOR
AUTOMATIC SCREW MACHINE
(c)
METRIC
Draw
half
size

A   B   C

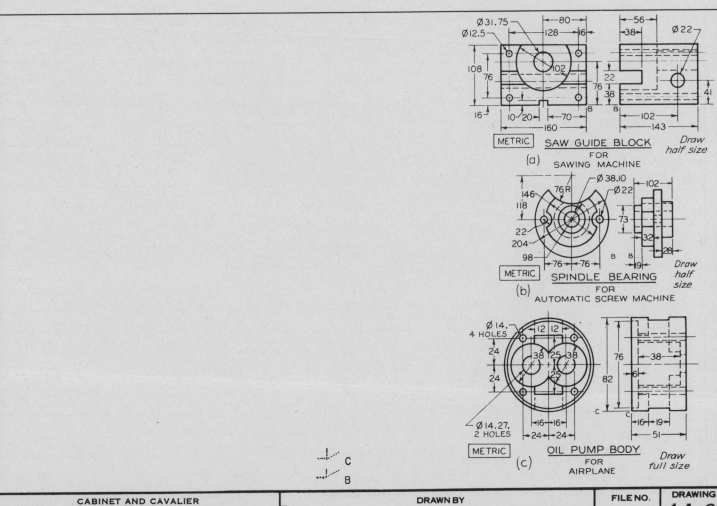

SAW GUIDE BLOCK
FOR
SAWING MACHINE
(a)
METRIC
Draw
half size

SPINDLE BEARING
FOR
AUTOMATIC SCREW MACHINE
(b)
METRIC
Draw
half
size

OIL PUMP BODY
FOR
AIRPLANE
(c)
METRIC
Draw
full size

C
B

| CABINET AND CAVALIER | DRAWN BY | FILE NO. | DRAWING |
| OBLIQUE PROJECTION | | | 14-2 |

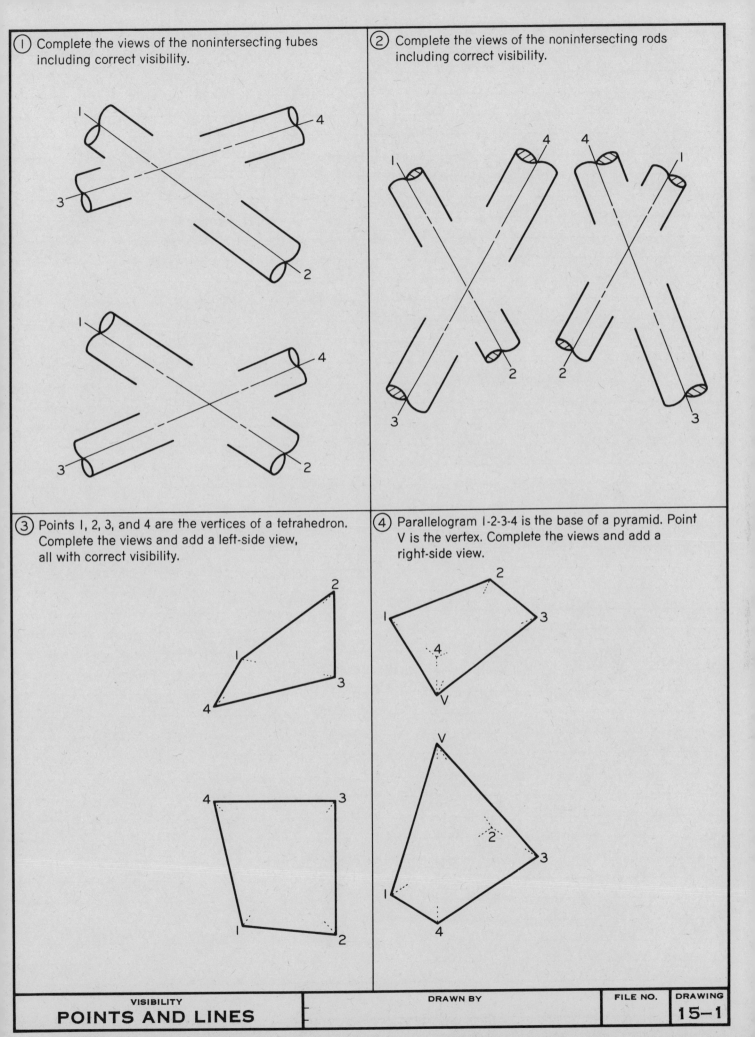

① Complete the views of the nonintersecting tubes including correct visibility.

② Complete the views of the nonintersecting rods including correct visibility.

③ Points 1, 2, 3, and 4 are the vertices of a tetrahedron. Complete the views and add a left-side view, all with correct visibility.

④ Parallelogram 1-2-3-4 is the base of a pyramid. Point V is the vertex. Complete the views and add a right-side view.

VISIBILITY
POINTS AND LINES

DRAWN BY

FILE NO.

DRAWING
15—1

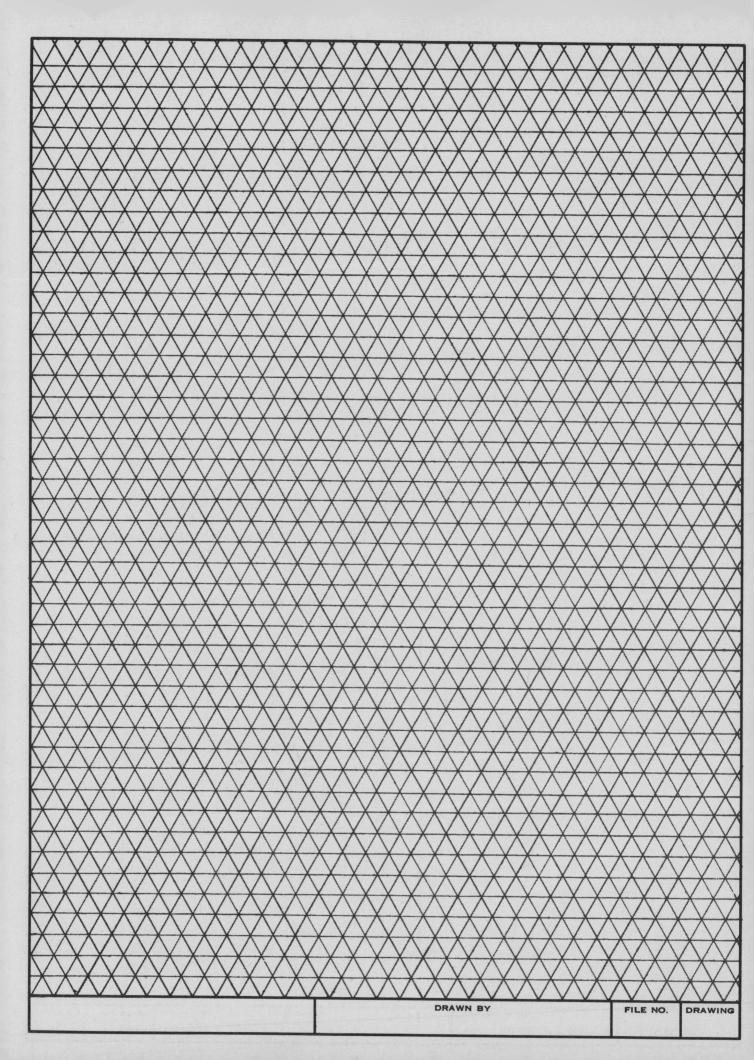

DRAWN BY

FILE NO.

DRAWING

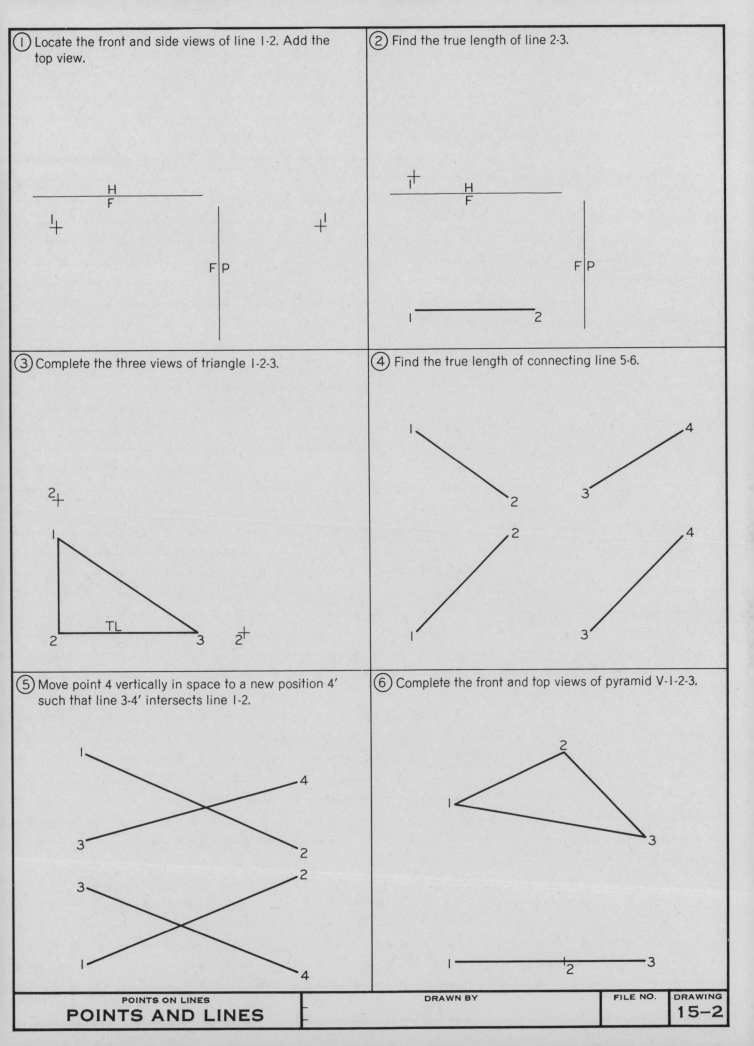

① Locate the front and side views of line 1-2. Add the top view.

H
F

F P

② Find the true length of line 2-3.

H
F

F P

1 —————————— 2

③ Complete the three views of triangle 1-2-3.

2

1

2   TL   3   2

④ Find the true length of connecting line 5-6.

1
2

4
3

2
1

4
3

⑤ Move point 4 vertically in space to a new position 4' such that line 3-4' intersects line 1-2.

1
4
3
2

3
2
1
4

⑥ Complete the front and top views of pyramid V-1-2-3.

2
1
3

1 ————— 2 ————— 3

POINTS ON LINES
POINTS AND LINES

DRAWN BY

FILE NO.

DRAWING
15-2

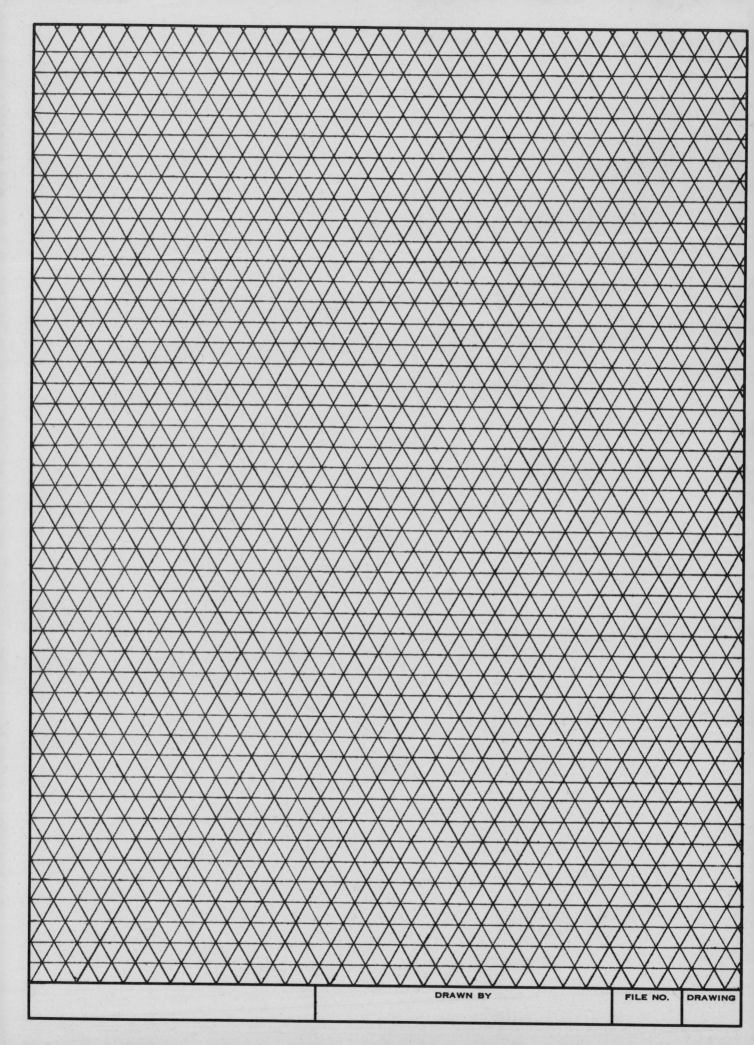

DRAWN BY   FILE NO.   DRAWING

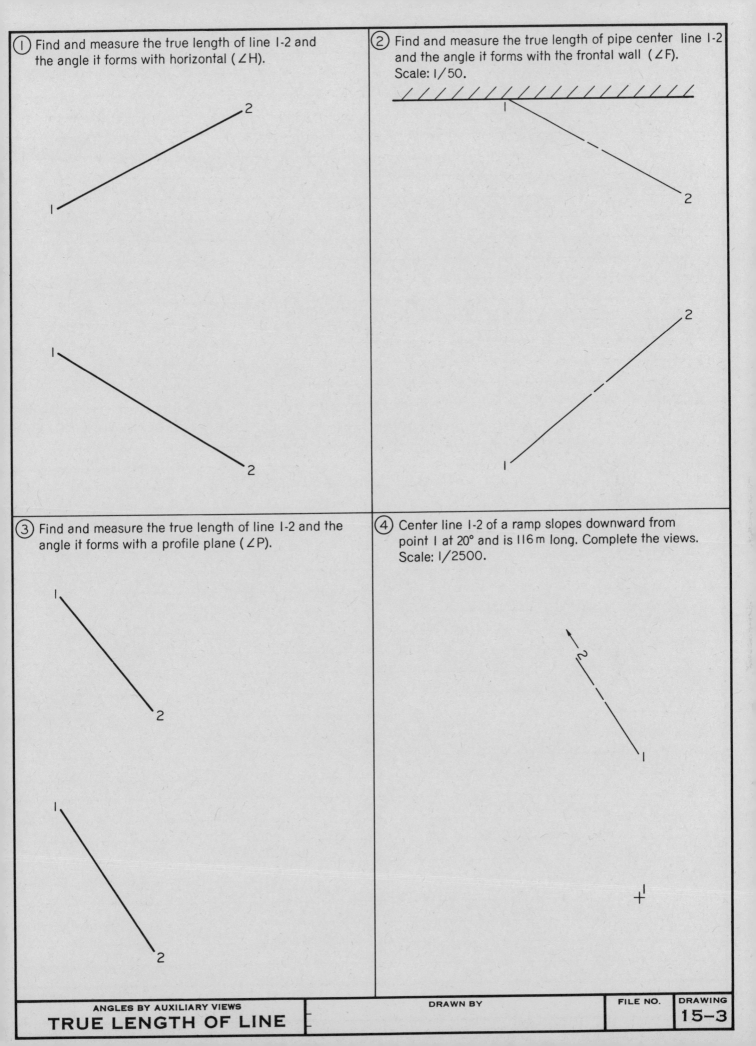

1. Find and measure the true length of line 1-2 and the angle it forms with horizontal (∠H).

2. Find and measure the true length of pipe center line 1-2 and the angle it forms with the frontal wall (∠F). Scale: 1/50.

3. Find and measure the true length of line 1-2 and the angle it forms with a profile plane (∠P).

4. Center line 1-2 of a ramp slopes downward from point 1 at 20° and is 116 m long. Complete the views. Scale: 1/2500.

ANGLES BY AUXILIARY VIEWS
TRUE LENGTH OF LINE

DRAWN BY

FILE NO.

DRAWING
15-3

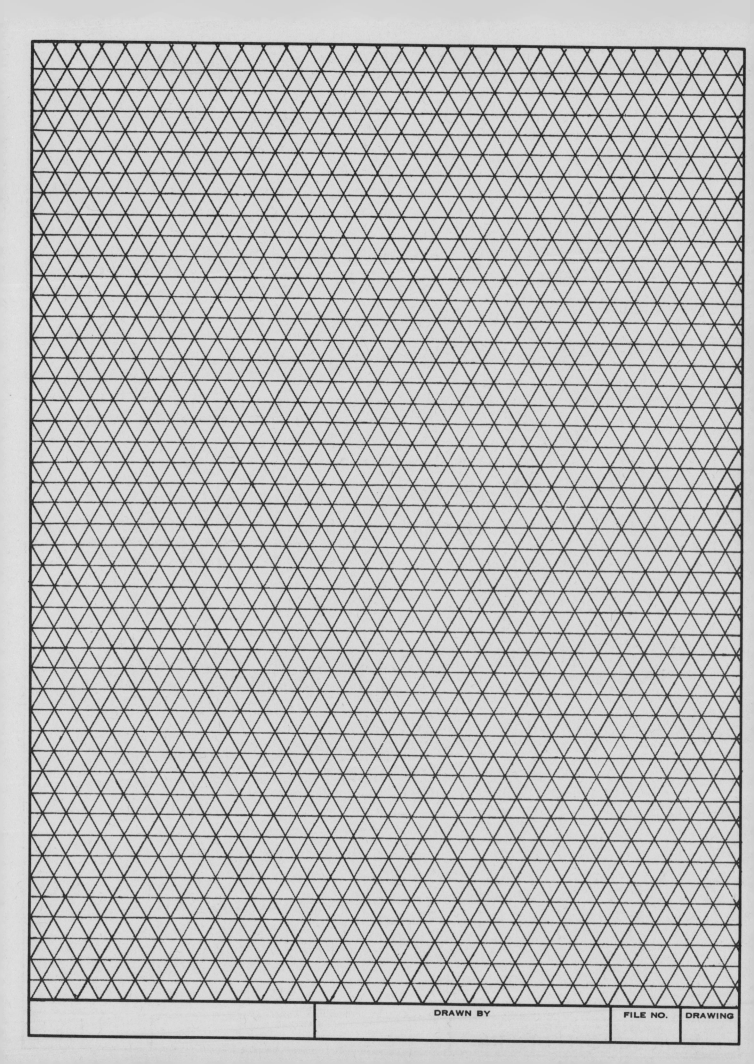

DRAWN BY   FILE NO.   DRAWING

① Measure the length, bearing, and grade of tunnel 1-2.
Scale: 1/5000.

② Measure the length, bearing, and grade of sluice 1-2.
Scale: 1/750.

③ A tunnel bears N40°E from point 1 at a grade of
−30%, to a point 2 that is 214 m along the tunnel.
Complete the views. Scale: 1/5000.

④ If cable 3-2 has the same grade as cable 1-2,
complete the front view. Measure the grade.

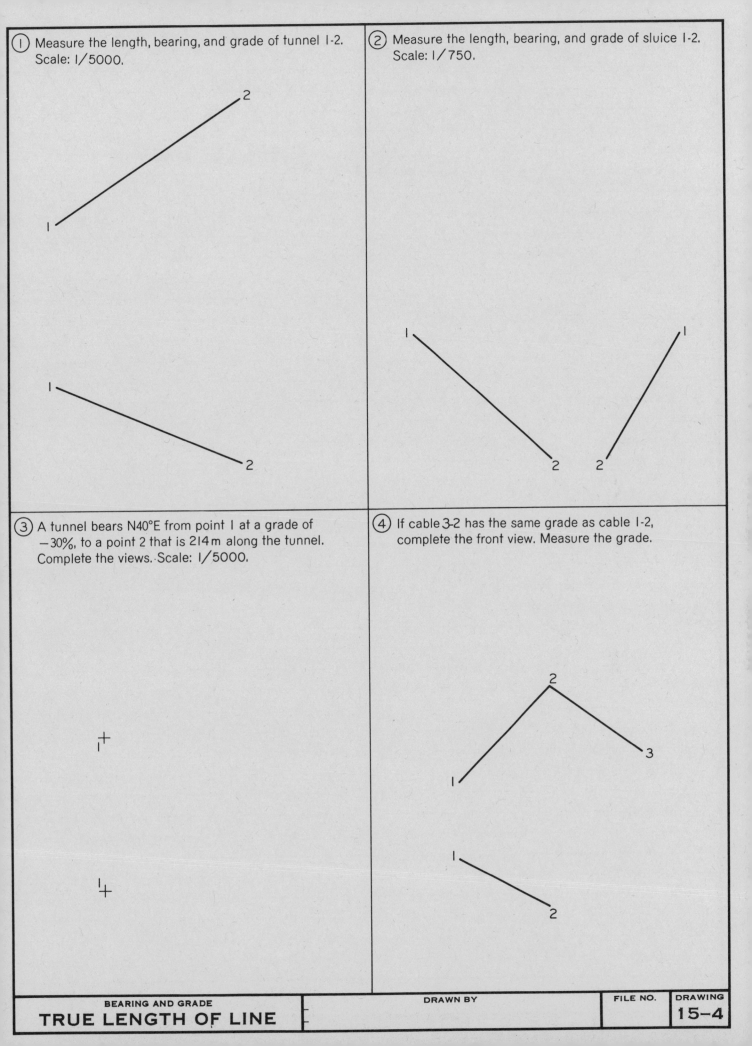

BEARING AND GRADE
TRUE LENGTH OF LINE

DRAWN BY

FILE NO.

DRAWING
15-4

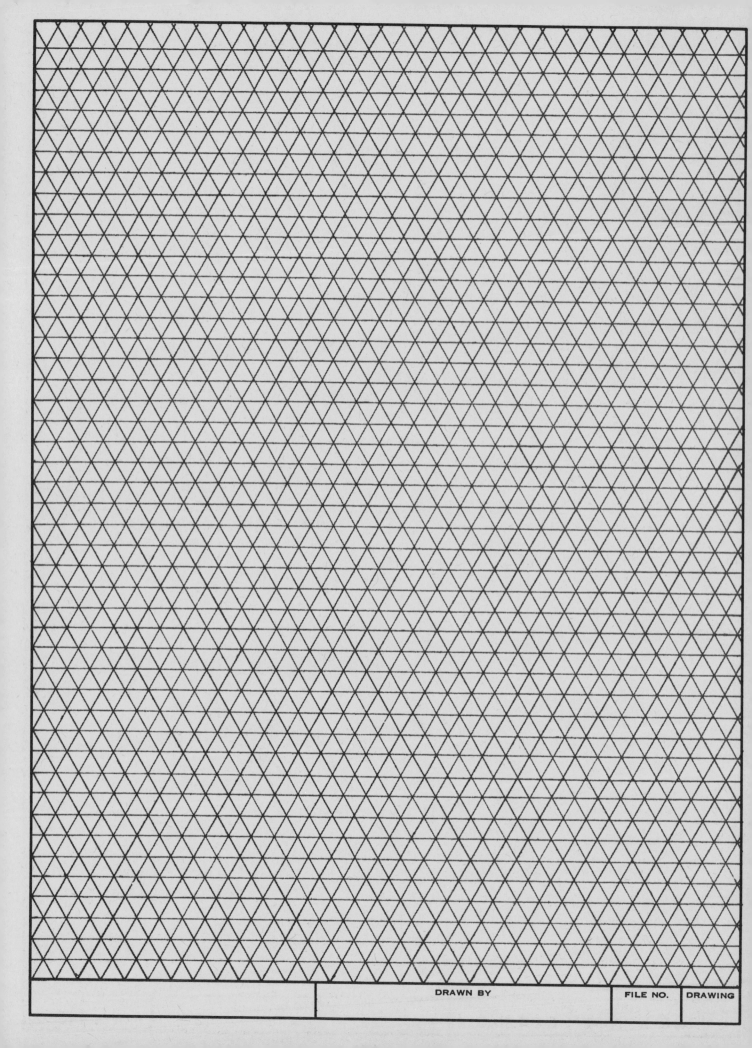

DRAWN BY
FILE NO.
DRAWING

① By revolution, find the true length and slope of line 1-2. Measure the bearing.

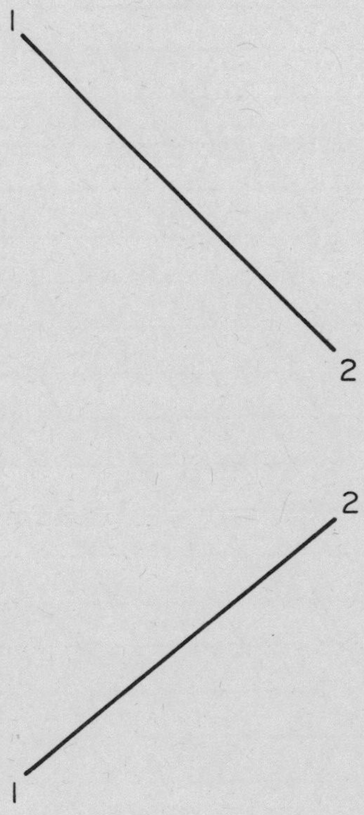

1

2

2

1

② Find the angles formed by line 1-2 and planes F and P. Measure the true length of line 1-2.

2

1

1

2

③ Line 1-2 forms an angle of 30° with a frontal plane. Complete its front view.

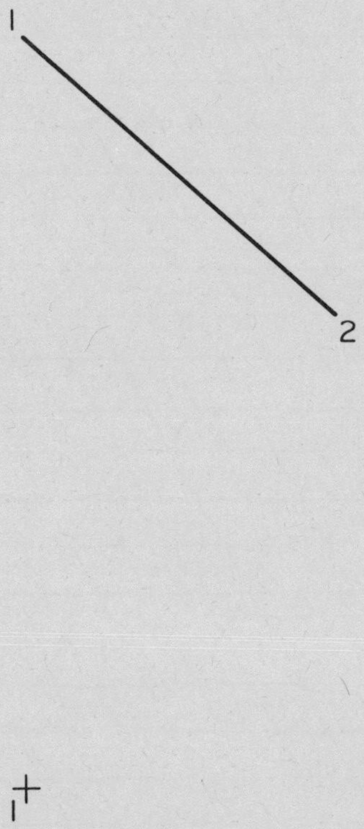

1

2

1+

④ An aircraft at point 1, on an azimuth course of N220°, is losing altitude at the rate of 400 m in 1000 m (map distance). Find the front and top views of a segment of the flight path.

+1

+1

REVOLUTION
TRUE LENGTH OF LINE

DRAWN BY

FILE NO.

DRAWING
15-5

DRAWN BY   FILE NO.   DRAWING

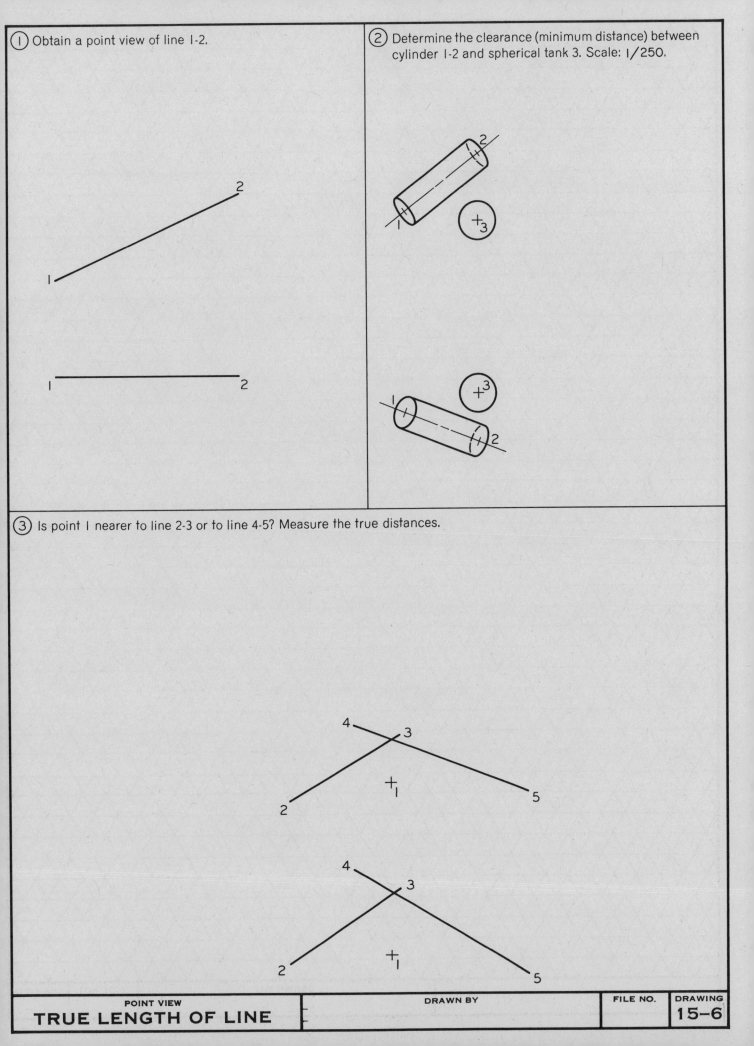

① Obtain a point view of line 1-2.

POINT VIEW

② Determine the clearance (minimum distance) between cylinder 1-2 and spherical tank 3. Scale: 1/250.

③ Is point 1 nearer to line 2-3 or to line 4-5? Measure the true distances.

POINT VIEW
TRUE LENGTH OF LINE

DRAWN BY

FILE NO.

DRAWING
15-6

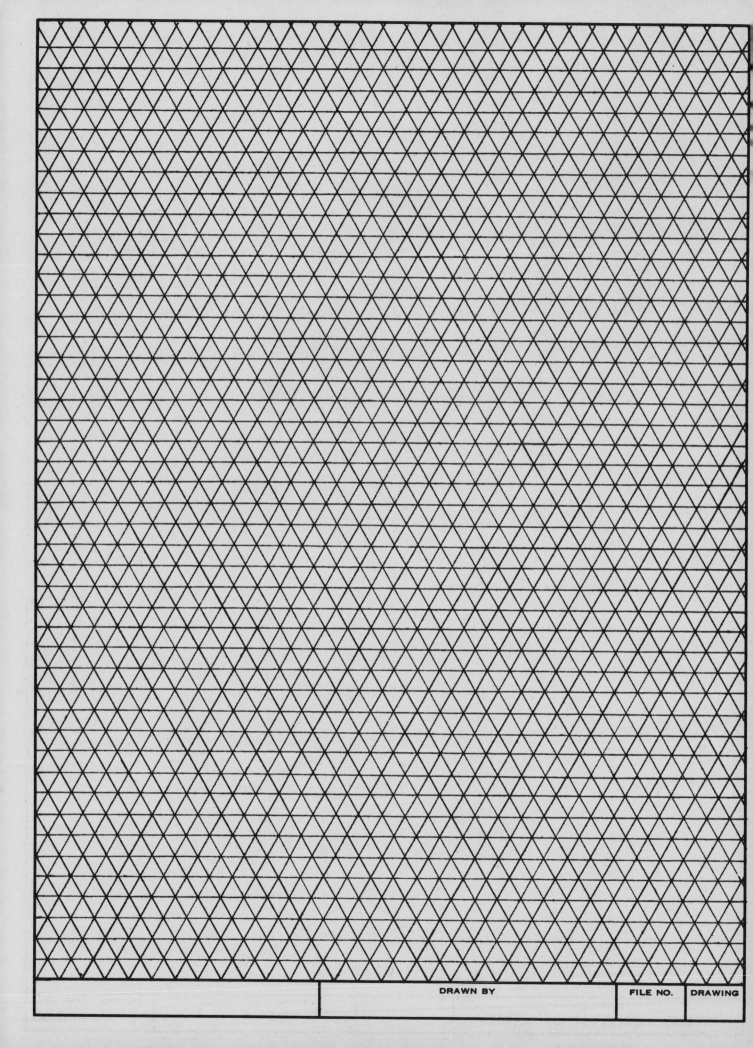

DRAWN BY | FILE NO. | DRAWING

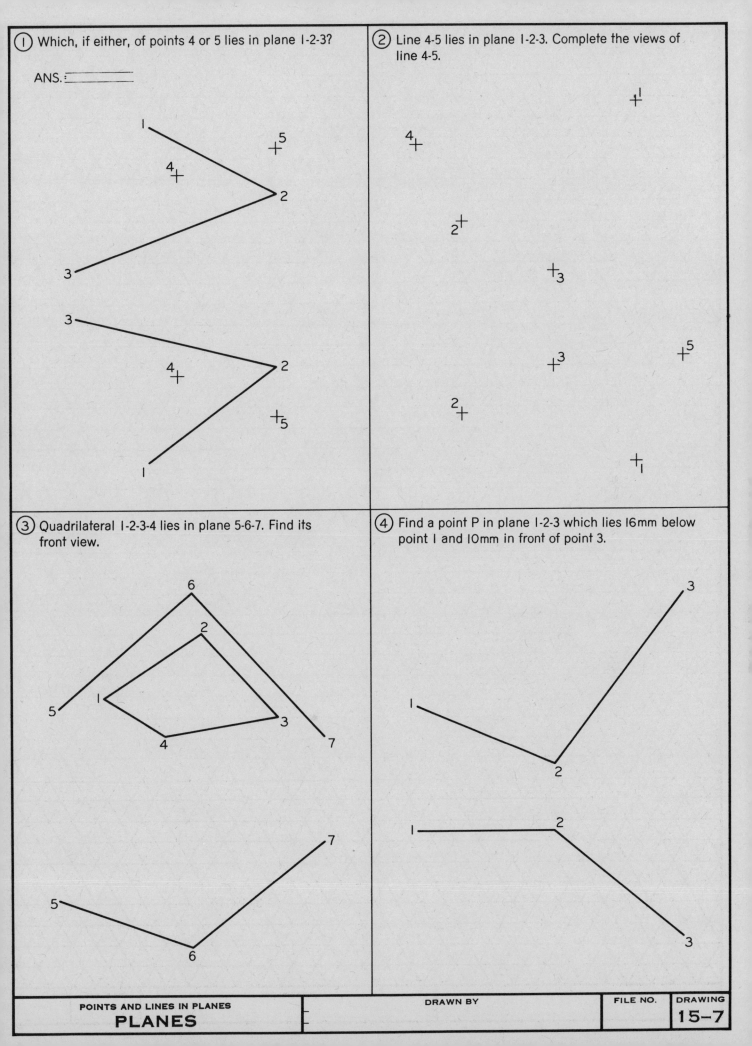

1. Which, if either, of points 4 or 5 lies in plane 1-2-3?

ANS. _____

2. Line 4-5 lies in plane 1-2-3. Complete the views of line 4-5.

3. Quadrilateral 1-2-3-4 lies in plane 5-6-7. Find its front view.

4. Find a point P in plane 1-2-3 which lies 16mm below point 1 and 10mm in front of point 3.

POINTS AND LINES IN PLANES

PLANES

DRAWN BY

FILE NO.

DRAWING

15-7

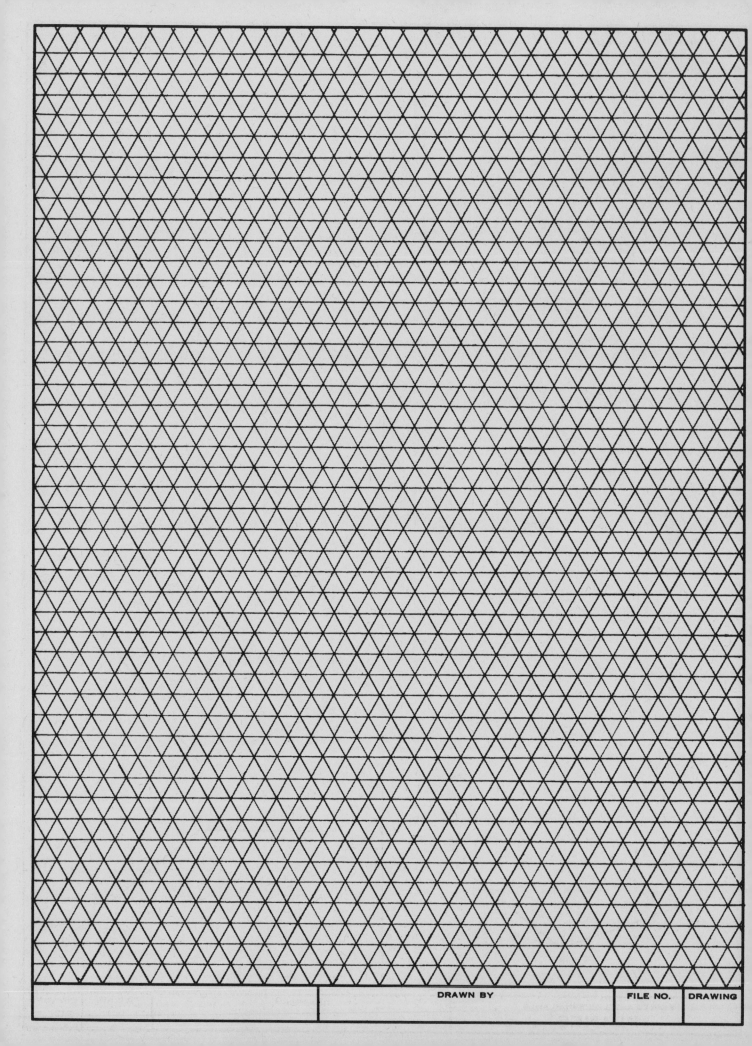

DRAWN BY

FILE NO.

DRAWING

① Construct a view showing the true size of triangle 1-2-3.
   Calculate the area of the triangle. Scale: 1/2500.

AREA = _____

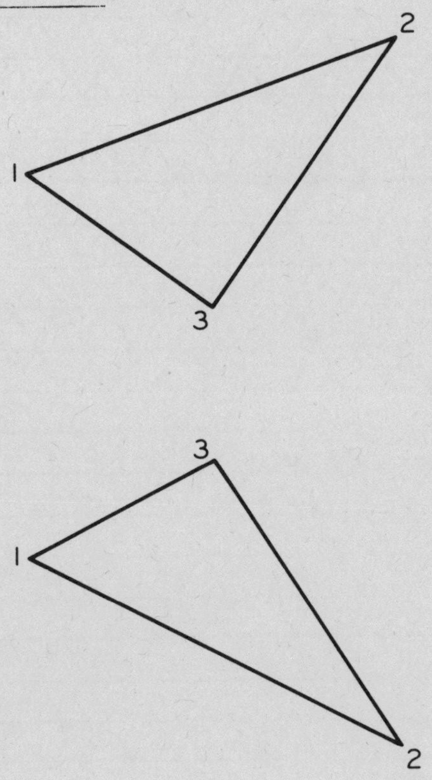

2

1

3

3

1

2

METRIC

② Pipe lines 1-2 and 3-4 are connected with a feeder branch using 45° lateral
   fittings. One fitting is located at point 5 on line 3-4.
   Find the views of the feeder branch.

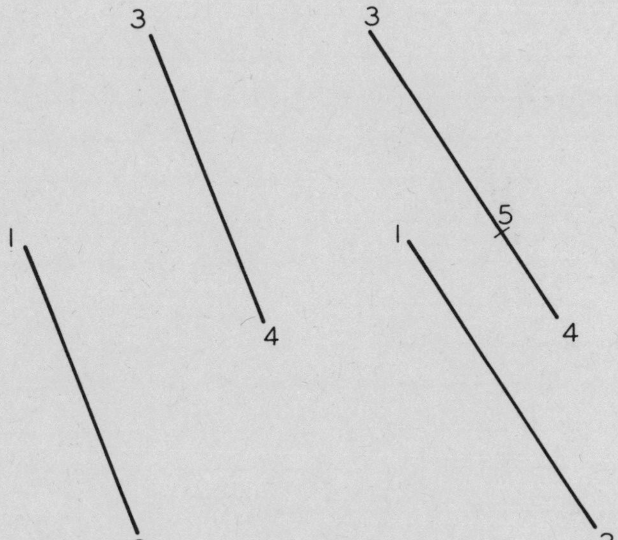

3

1

4

2

3

1

5

4

2

DRAWN BY

FILE NO.

DRAWING

TRUE SIZE
PLANES

15-8

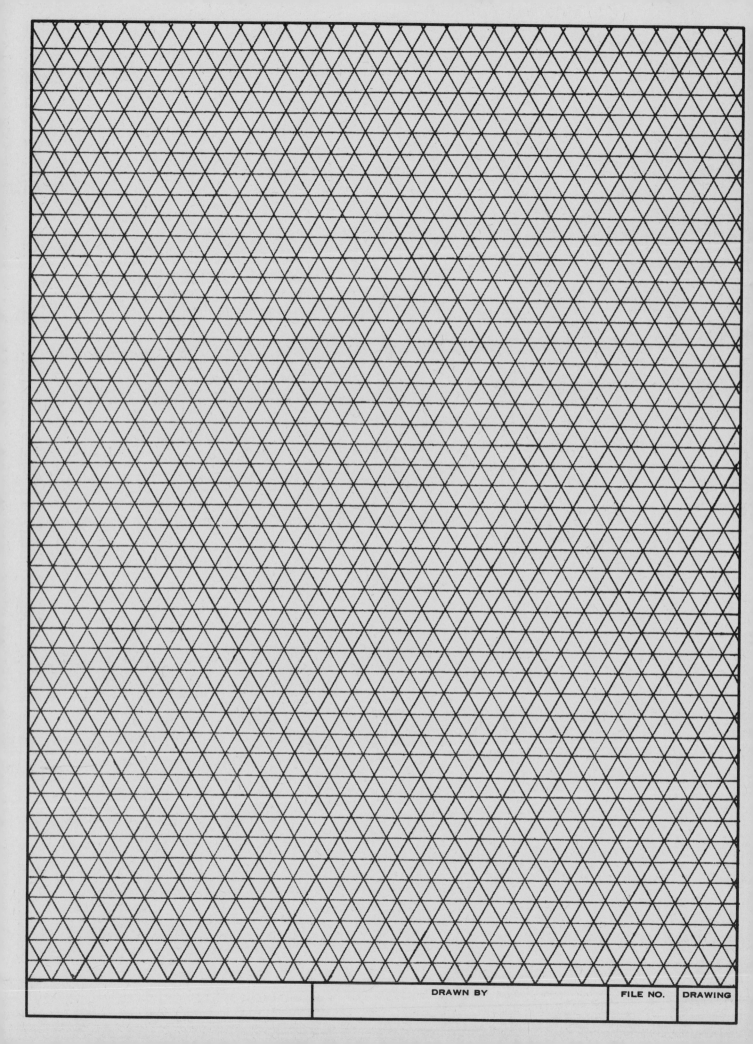

DRAWN BY

FILE NO.

DRAWING

① Find the piercing point of line 1-2 in plane 3-4-5.

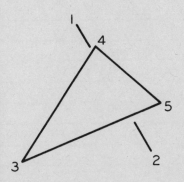

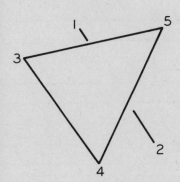

② Establish the piercing point of laser beam 1-2 in plane 3-4-5-6.

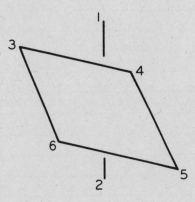

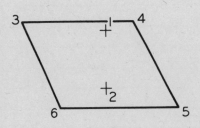

③ Show the piercing points of line 1-2 in the surfaces of the pyramid.

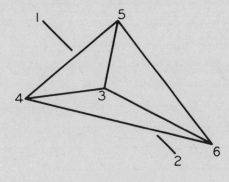

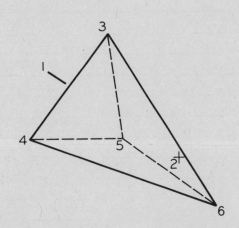

PIERCING POINTS—EDGE-VIEW METHOD
LINES AND PLANES

DRAWN BY

FILE NO.

DRAWING

15—9

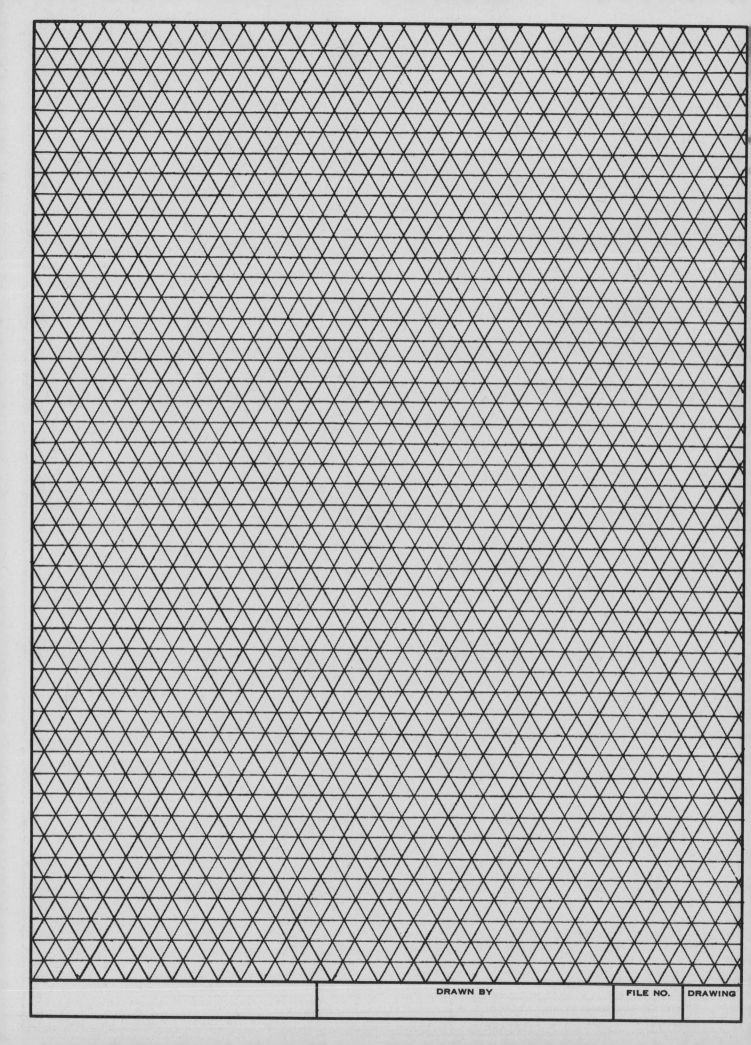

DRAWN BY | FILE NO. | DRAWING

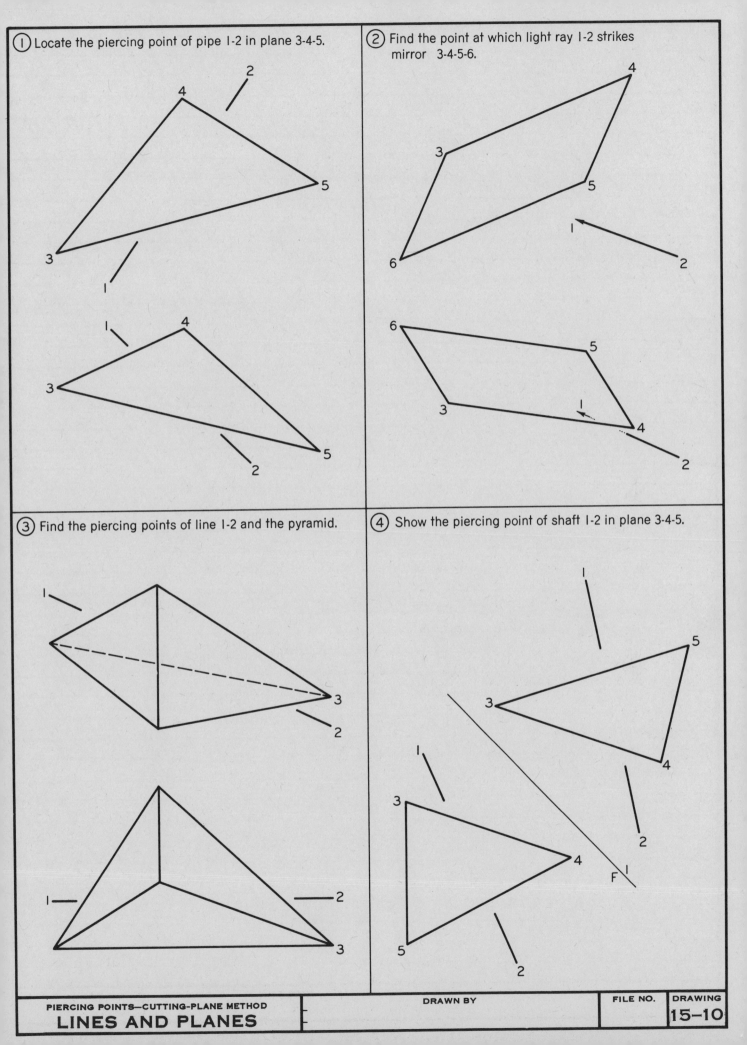

① Locate the piercing point of pipe 1-2 in plane 3-4-5.

② Find the point at which light ray 1-2 strikes mirror 3-4-5-6.

③ Find the piercing points of line 1-2 and the pyramid.

④ Show the piercing point of shaft 1-2 in plane 3-4-5.

PIERCING POINTS—CUTTING-PLANE METHOD
LINES AND PLANES

DRAWN BY

FILE NO.

DRAWING
15–10

DRAWN BY

FILE NO. DRAWING

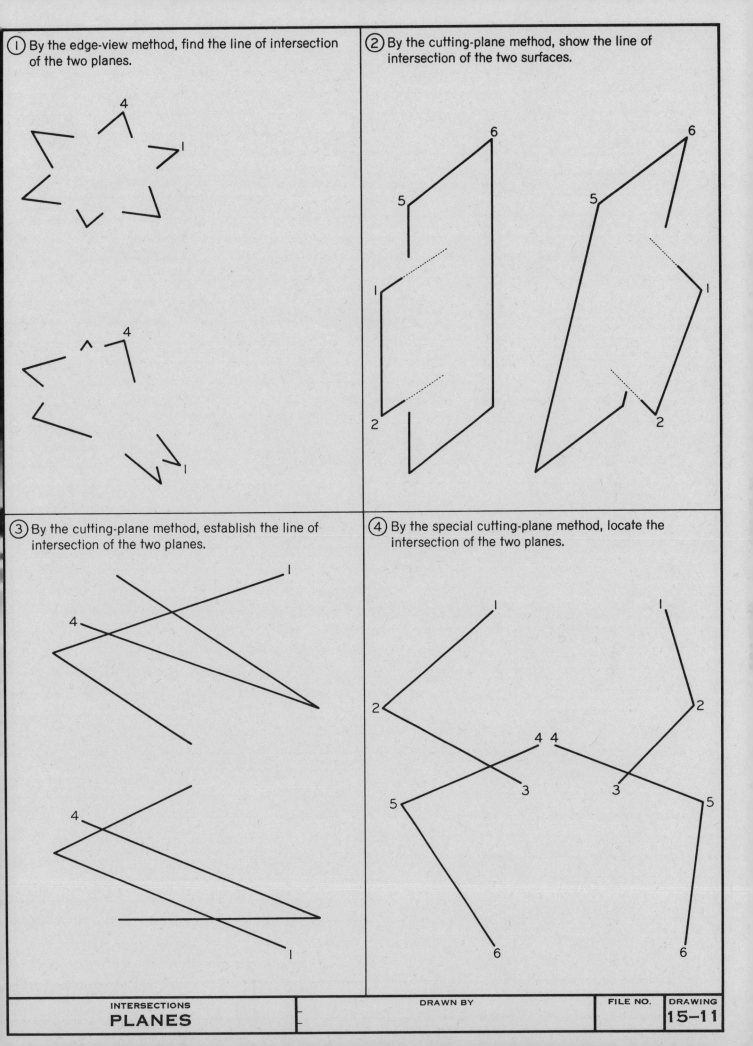

1. By the edge-view method, find the line of intersection of the two planes.

2. By the cutting-plane method, show the line of intersection of the two surfaces.

3. By the cutting-plane method, establish the line of intersection of the two planes.

4. By the special cutting-plane method, locate the intersection of the two planes.

INTERSECTIONS
PLANES

DRAWN BY

FILE NO.

DRAWING
15-11

DRAWN BY   FILE NO.   DRAWING

① Find the dihedral angles between the lateral faces of the prism.

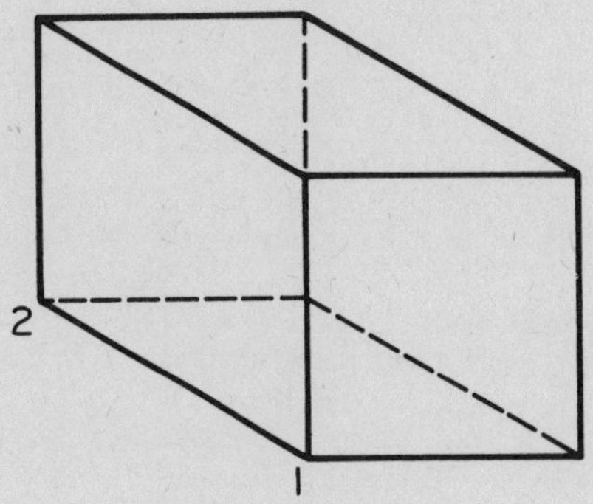

2

1

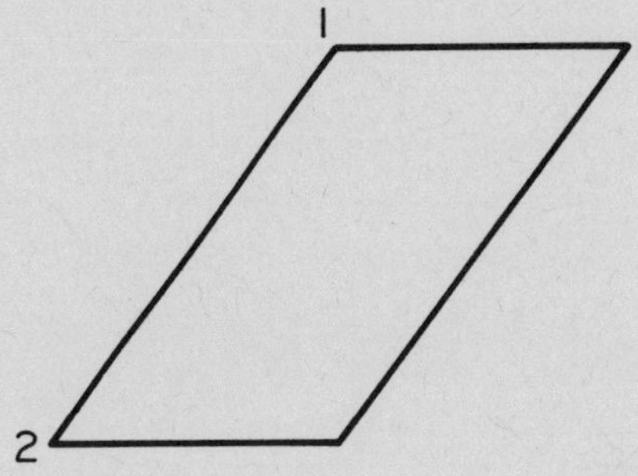

1

2

② Determine the angle between the roof planes.

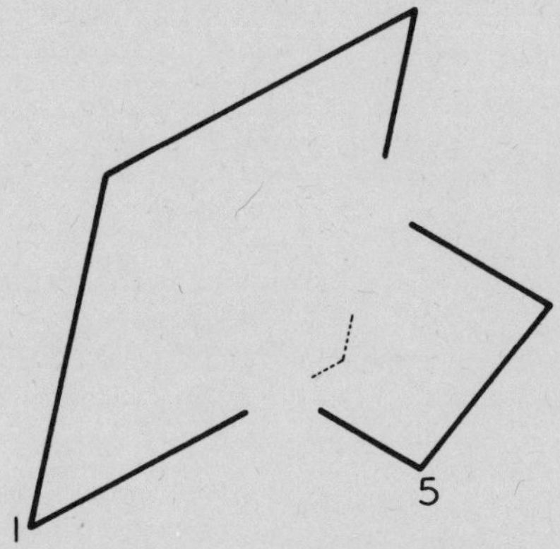

5

1

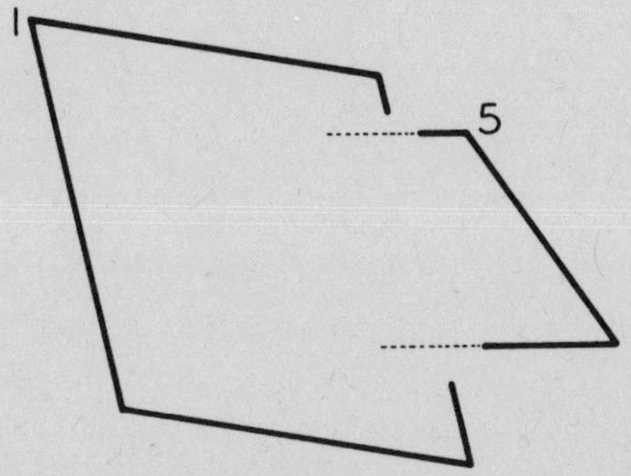

1

5

DIHEDRAL ANGLES
PLANES

DRAWN BY

FILE NO.

DRAWING
15–12

DRAWN BY   FILE NO.   DRAWING

① Find the angle between control cable 1-2 and
bulkhead 3-4-5-6.

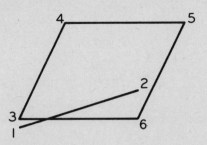

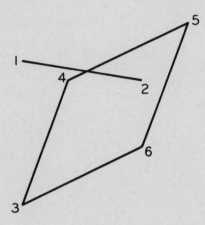

② Establish the views of 38 mm line 1-2 such that the line 1-2 forms
an angle of 25° with the given surface.

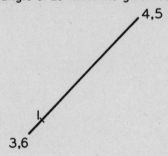

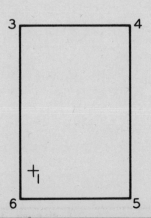

ANGLE BETWEEN LINE AND PLANE	DRAWN BY	FILE NO.	DRAWING
LINES AND PLANES			**15–13**

DRAWN BY

FILE NO.

DRAWING

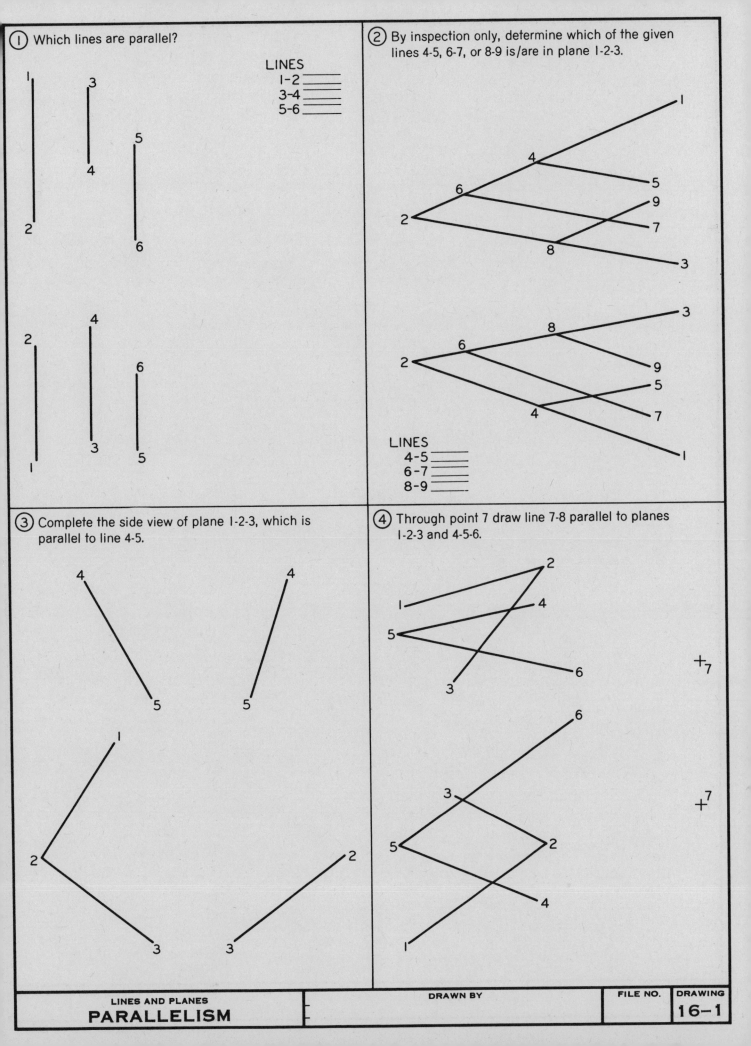

① Which lines are parallel?

LINES
1-2 ————
3-4 ————
5-6 ————

② By inspection only, determine which of the given lines 4-5, 6-7, or 8-9 is/are in plane 1-2-3.

LINES
4-5 ————
6-7 ————
8-9 ————

③ Complete the side view of plane 1-2-3, which is parallel to line 4-5.

④ Through point 7 draw line 7-8 parallel to planes 1-2-3 and 4-5-6.

LINES AND PLANES
PARALLELISM

DRAWN BY

FILE NO.

DRAWING
16-1

DRAWN BY   FILE NO.   DRAWING

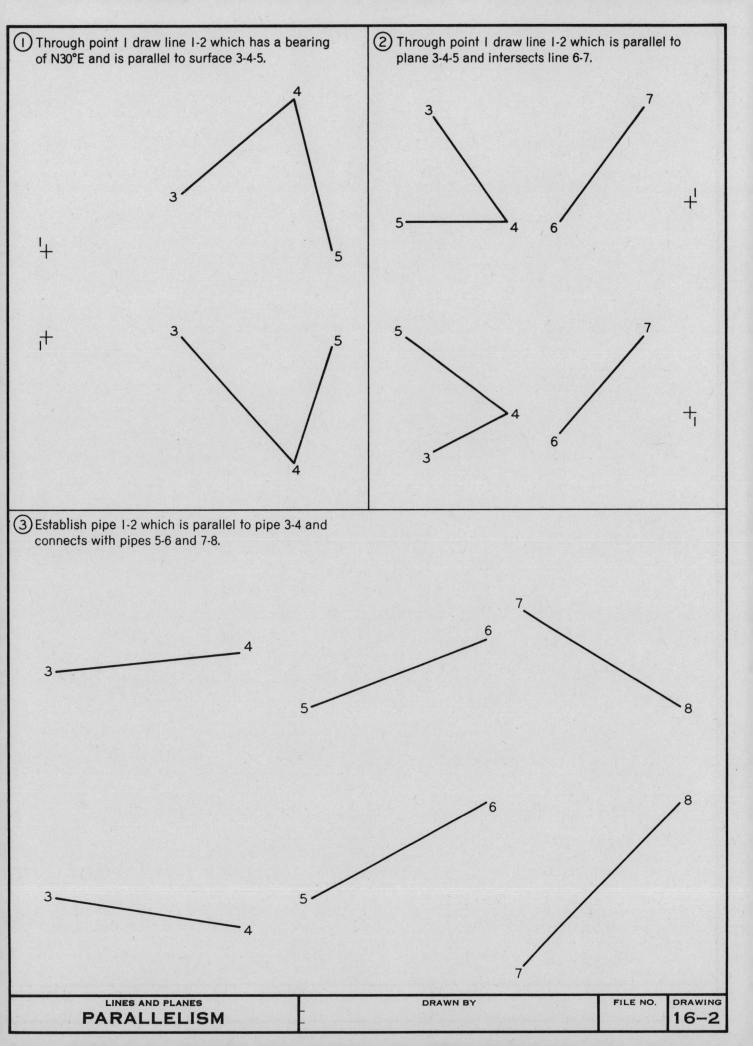

1. Through point I draw line I-2 which has a bearing of N30°E and is parallel to surface 3-4-5.

2. Through point I draw line I-2 which is parallel to plane 3-4-5 and intersects line 6-7.

3. Establish pipe I-2 which is parallel to pipe 3-4 and connects with pipes 5-6 and 7-8.

LINES AND PLANES
PARALLELISM

DRAWN BY

FILE NO.

DRAWING
16-2

DRAWN BY | FILE NO. | DRAWING

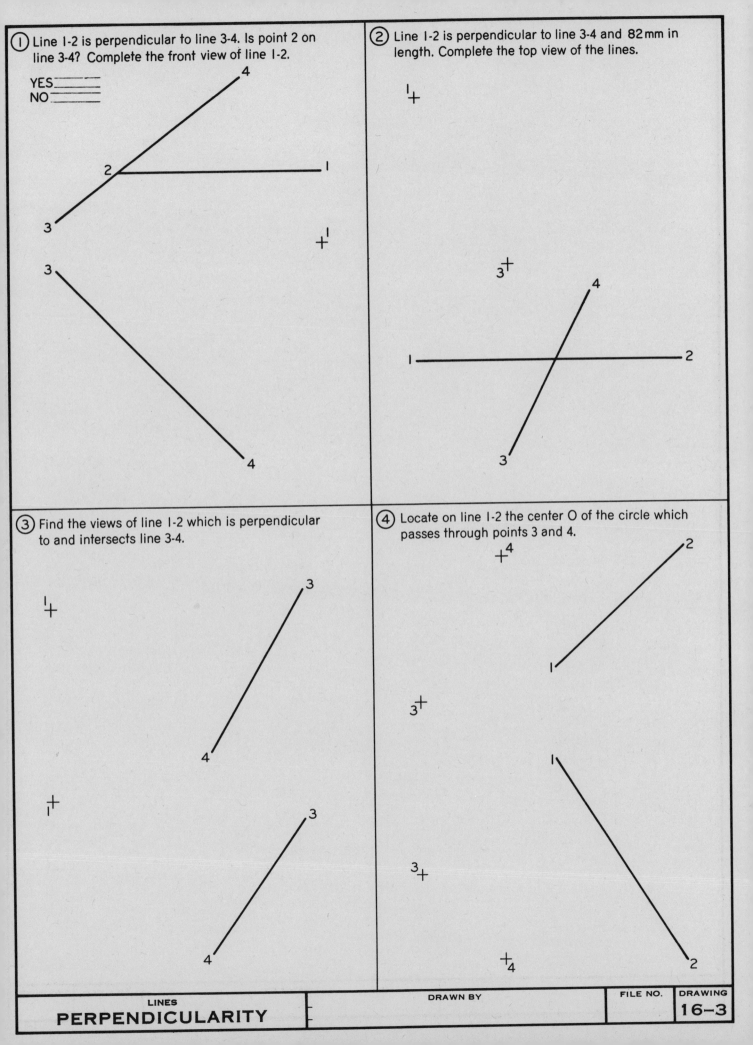

① Line 1-2 is perpendicular to line 3-4. Is point 2 on line 3-4? Complete the front view of line 1-2.

YES ___
NO ___

② Line 1-2 is perpendicular to line 3-4 and 82 mm in length. Complete the top view of the lines.

③ Find the views of line 1-2 which is perpendicular to and intersects line 3-4.

④ Locate on line 1-2 the center O of the circle which passes through points 3 and 4.

LINES
PERPENDICULARITY

DRAWN BY

FILE NO.

DRAWING
16-3

DRAWN BY

FILE NO.

DRAWING

① Locate the shadow of point 1 on surface 2-3-4-5 if the light rays are perpendicular to the surface.

② Measure the length and show the views of the altitude of all cones having point V as the vertex and with their bases in plane 1-2-3.

③ The axis of a right pyramid lies along line 1-2. The vertex is at point V. The base is an equilateral triangle with one corner at point 3. Complete all views.

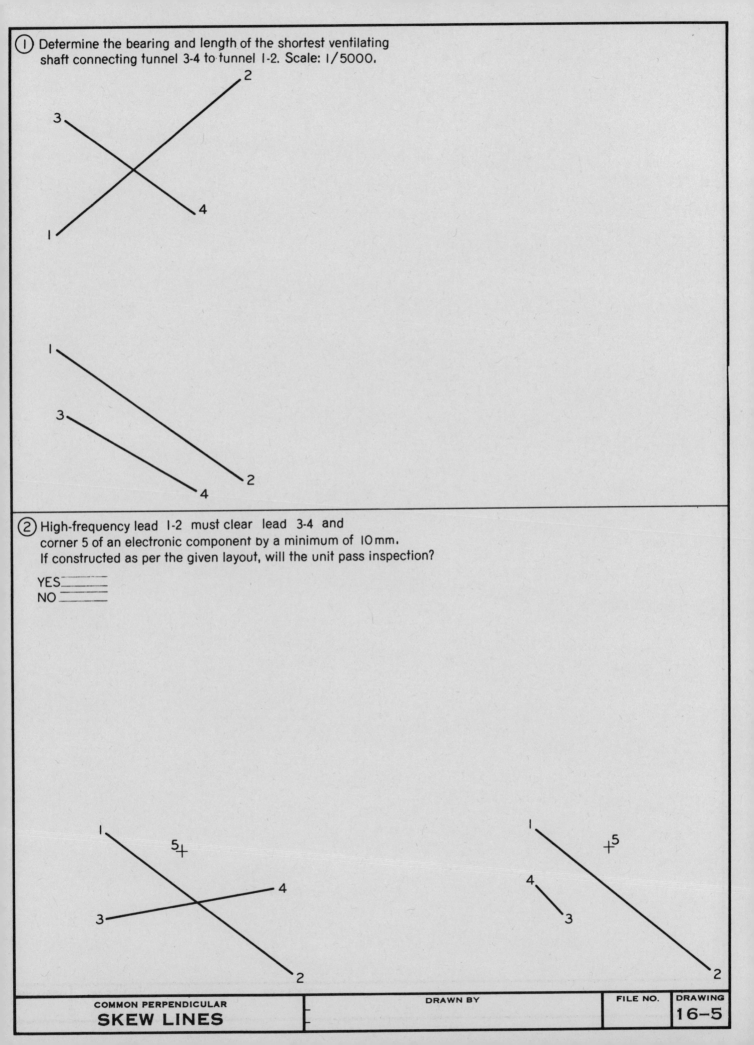

① Determine the bearing and length of the shortest ventilating
shaft connecting tunnel 3-4 to tunnel 1-2. Scale: 1/5000.

② High-frequency lead 1-2 must clear lead 3-4 and
corner 5 of an electronic component by a minimum of 10 mm.
If constructed as per the given layout, will the unit pass inspection?

YES ══════
NO ══════

| COMMON PERPENDICULAR | DRAWN BY | FILE NO. | DRAWING |
| **SKEW LINES** | | | 16-5 |

① Connect pipes 1-2 and 3-4 with the shortest branch parallel to the side (profile) wall.
Determine the length and show the views of the branch.
Scale: 1/100.

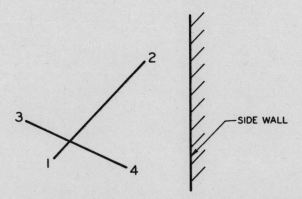

SIDE WALL

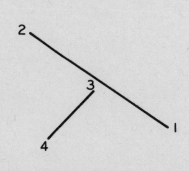

② Ski slope 1-2 is connected to ski slope 3-4 with the shortest path having a grade of − 10%.
Find the length and bearing and show the views of the path.
Scale: 1/5000.

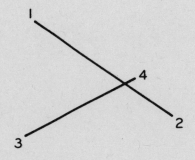

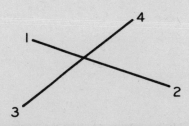

LINES AT SPECIFIED ANGLES
SKEW LINES

DRAWN BY

FILE NO.

DRAWING
16−6

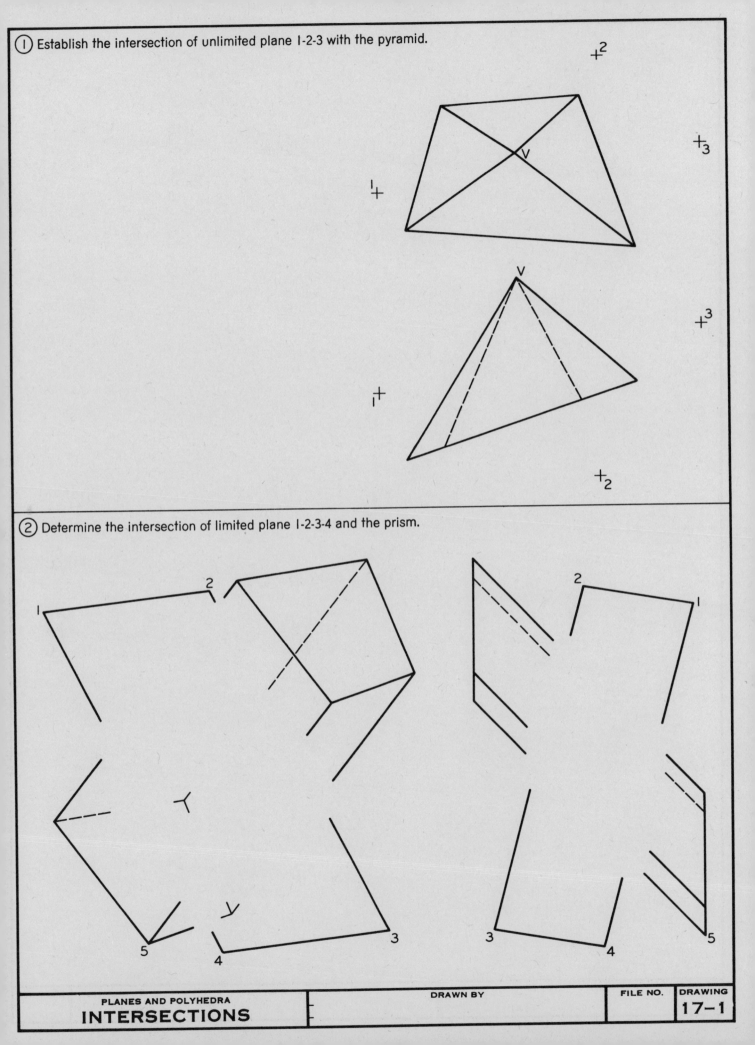

① Establish the intersection of unlimited plane 1-2-3 with the pyramid.

② Determine the intersection of limited plane 1-2-3-4 and the prism.

PLANES AND POLYHEDRA
INTERSECTIONS

DRAWN BY

FILE NO.

DRAWING
17—1

DRAWN BY   FILE NO.   DRAWING

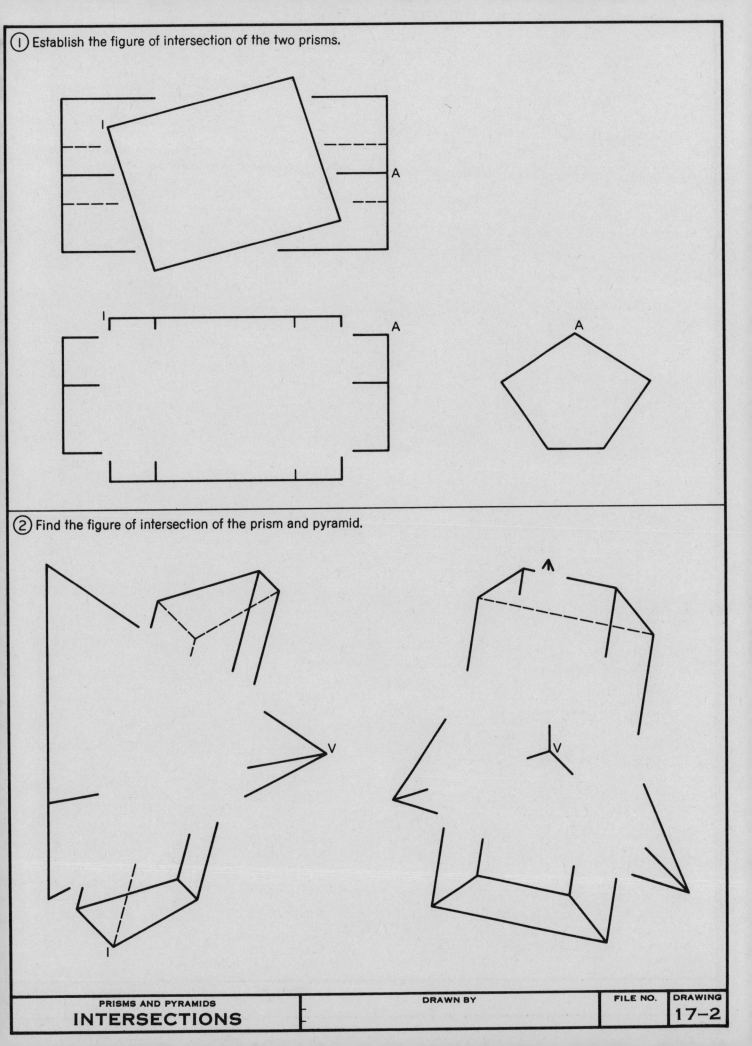

① Establish the figure of intersection of the two prisms.

② Find the figure of intersection of the prism and pyramid.

PRISMS AND PYRAMIDS
INTERSECTIONS

DRAWN BY

FILE NO.

DRAWING
17-2

DRAWN BY | FILE NO. | DRAWING

Complete the views of the intersecting forms of the Collector.

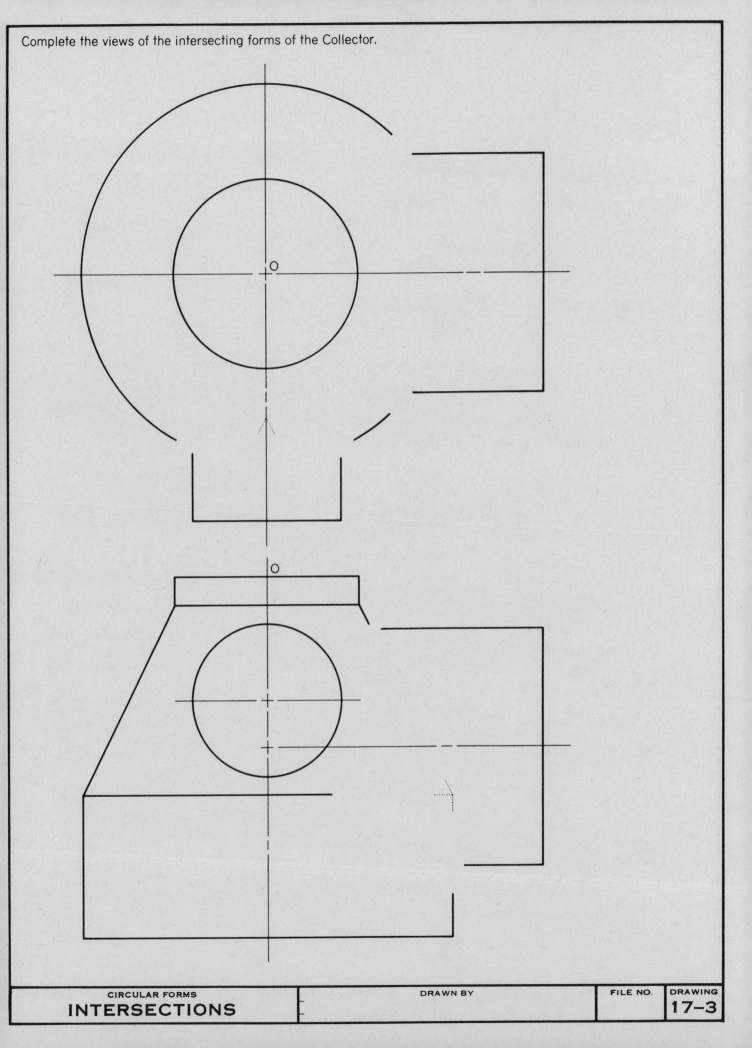

CIRCULAR FORMS
INTERSECTIONS

DRAWN BY

FILE NO.

DRAWING
17-3

DRAWN BY   FILE NO.   DRAWING

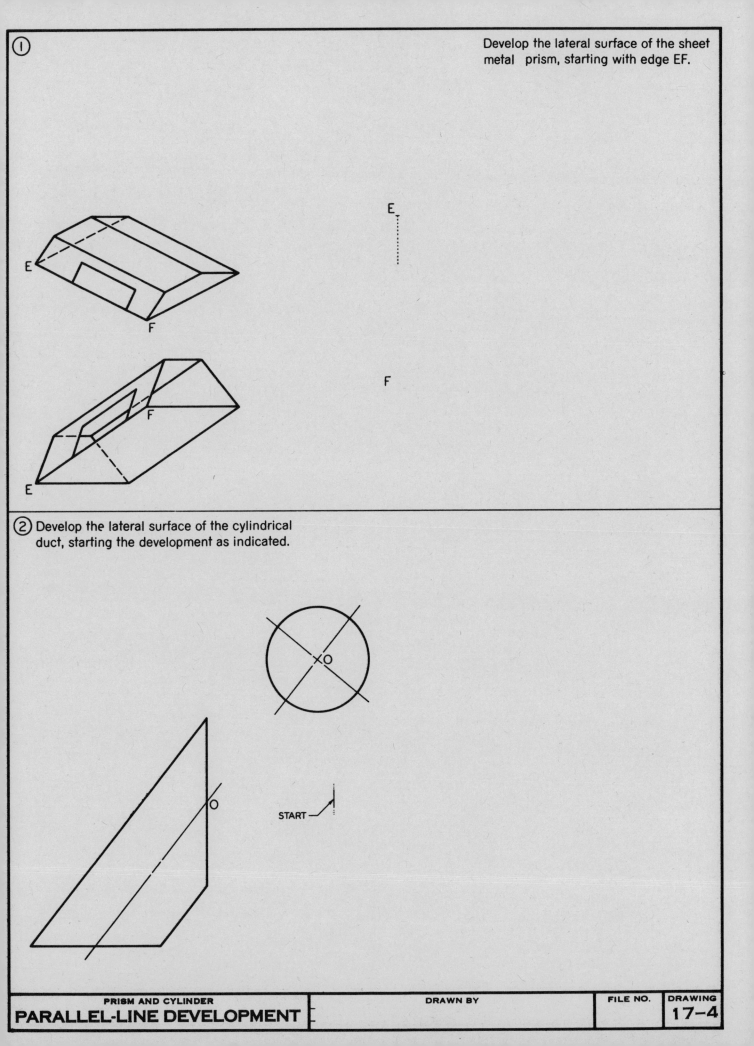

① Develop the lateral surface of the sheet metal prism, starting with edge EF.

E

F

② Develop the lateral surface of the cylindrical duct, starting the development as indicated.

O

START

O

PRISM AND CYLINDER
PARALLEL-LINE DEVELOPMENT

DRAWN BY

FILE NO.

DRAWING
17-4

DRAWN BY | FILE NO. | DRAWING

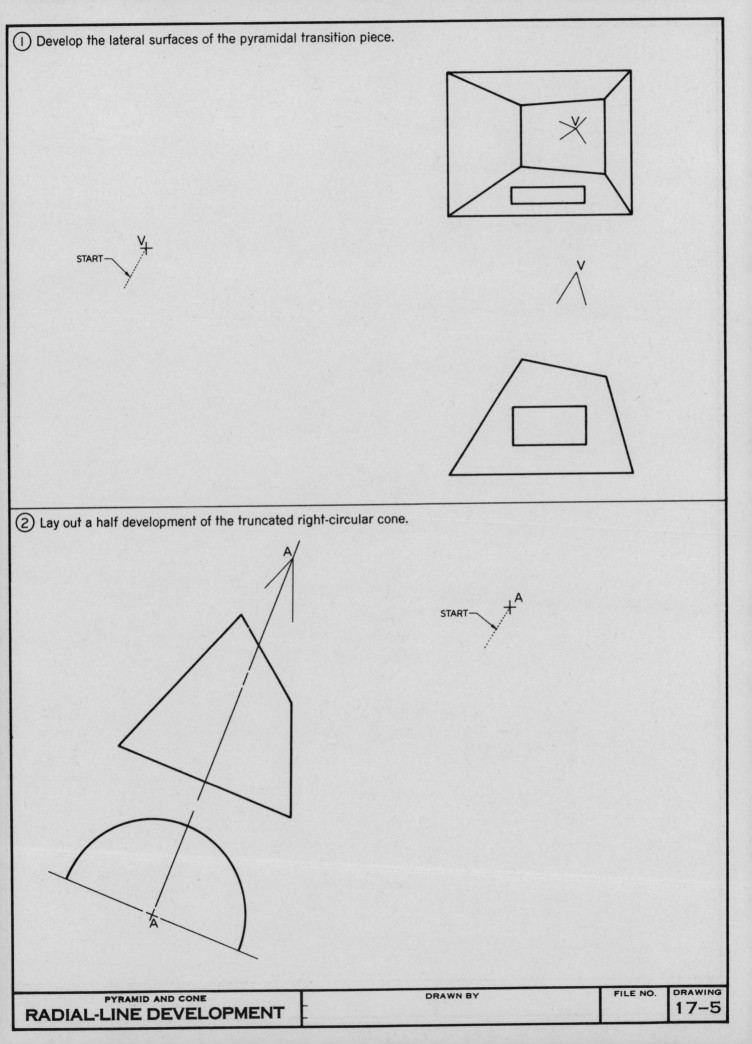

① Develop the lateral surfaces of the pyramidal transition piece.

V

START

V

② Lay out a half development of the truncated right-circular cone.

A

START A

A

PYRAMID AND CONE
RADIAL-LINE DEVELOPMENT

DRAWN BY

FILE NO.

DRAWING
17–5

Construct a half development of the transition piece, starting with seam E-1 at the indicated position and ending at the center line F-7 of panel B-C-7.

START

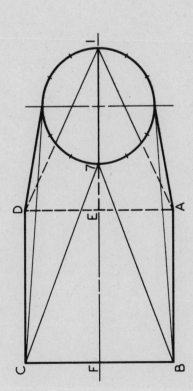

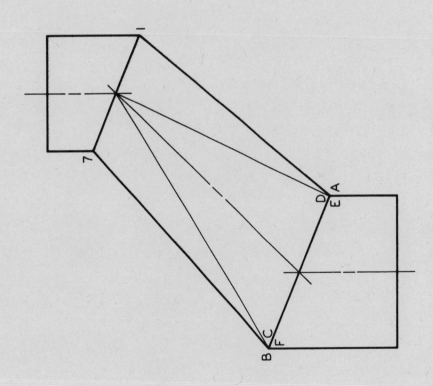

DRAWN BY | FILE NO. | DRAWING

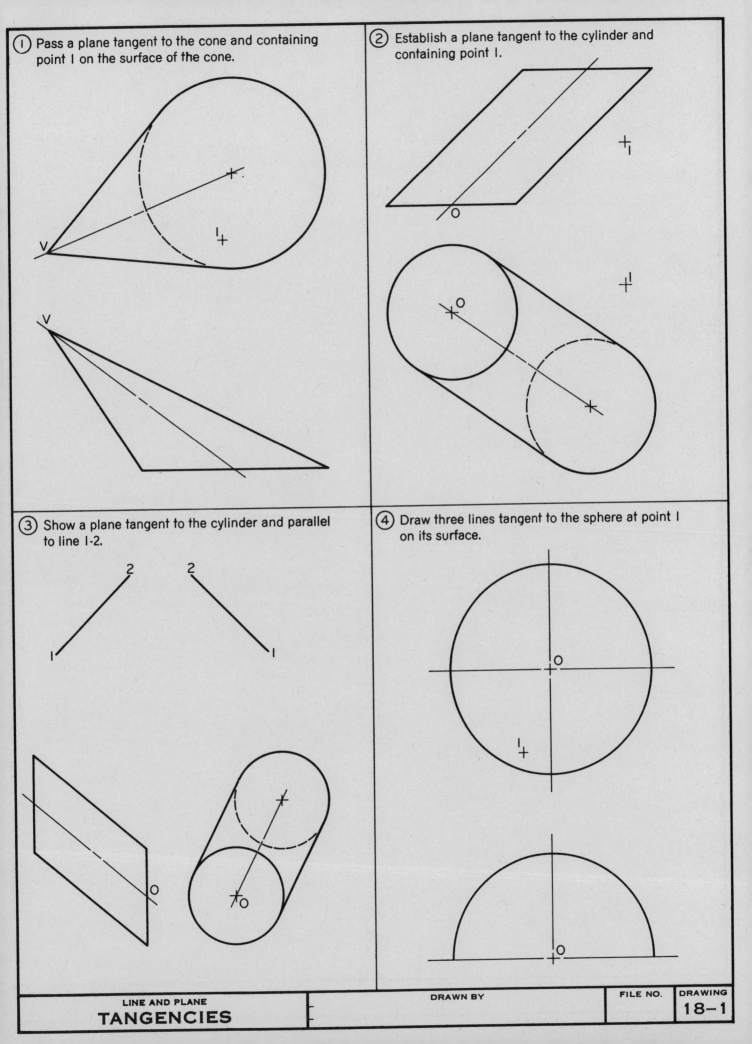

① Pass a plane tangent to the cone and containing point 1 on the surface of the cone.

② Establish a plane tangent to the cylinder and containing point 1.

③ Show a plane tangent to the cylinder and parallel to line 1-2.

④ Draw three lines tangent to the sphere at point 1 on its surface.

LINE AND PLANE
TANGENCIES

DRAWN BY

FILE NO.

DRAWING
18-1

DRAWN BY | FILE NO. | DRAWING

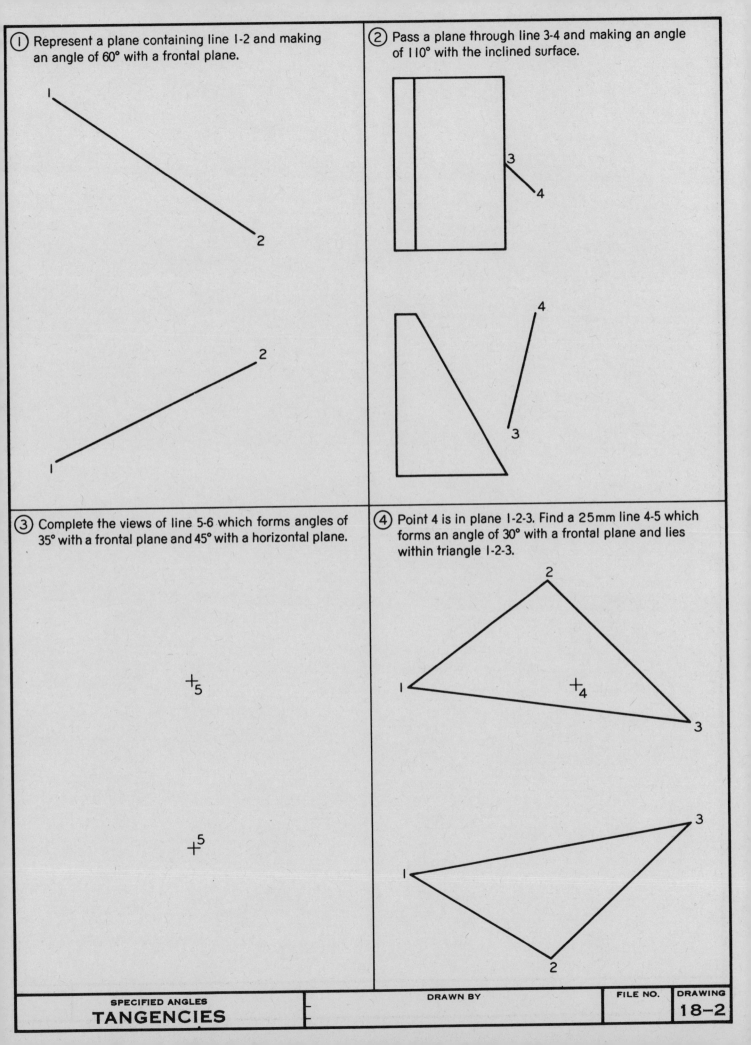

① Represent a plane containing line 1-2 and making an angle of 60° with a frontal plane.

② Pass a plane through line 3-4 and making an angle of 110° with the inclined surface.

③ Complete the views of line 5-6 which forms angles of 35° with a frontal plane and 45° with a horizontal plane.

④ Point 4 is in plane 1-2-3. Find a 25mm line 4-5 which forms an angle of 30° with a frontal plane and lies within triangle 1-2-3.

SPECIFIED ANGLES
TANGENCIES

DRAWN BY

FILE NO.

DRAWING
18-2

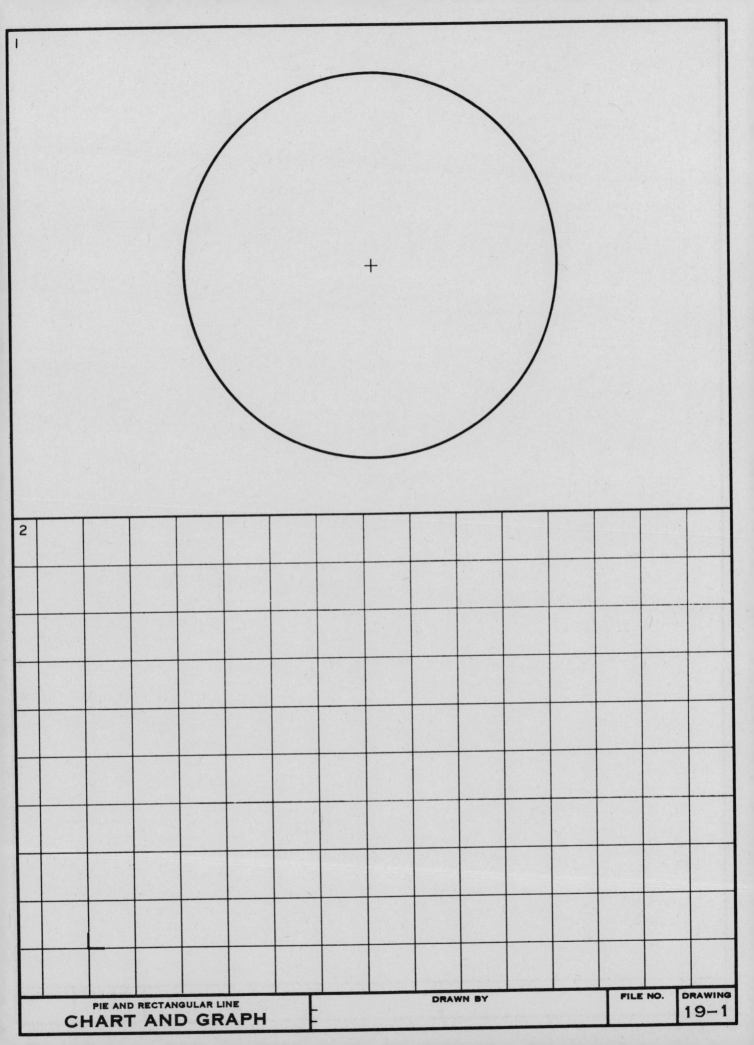

1

2

DRAWN BY

FILE NO.

DRAWING

①

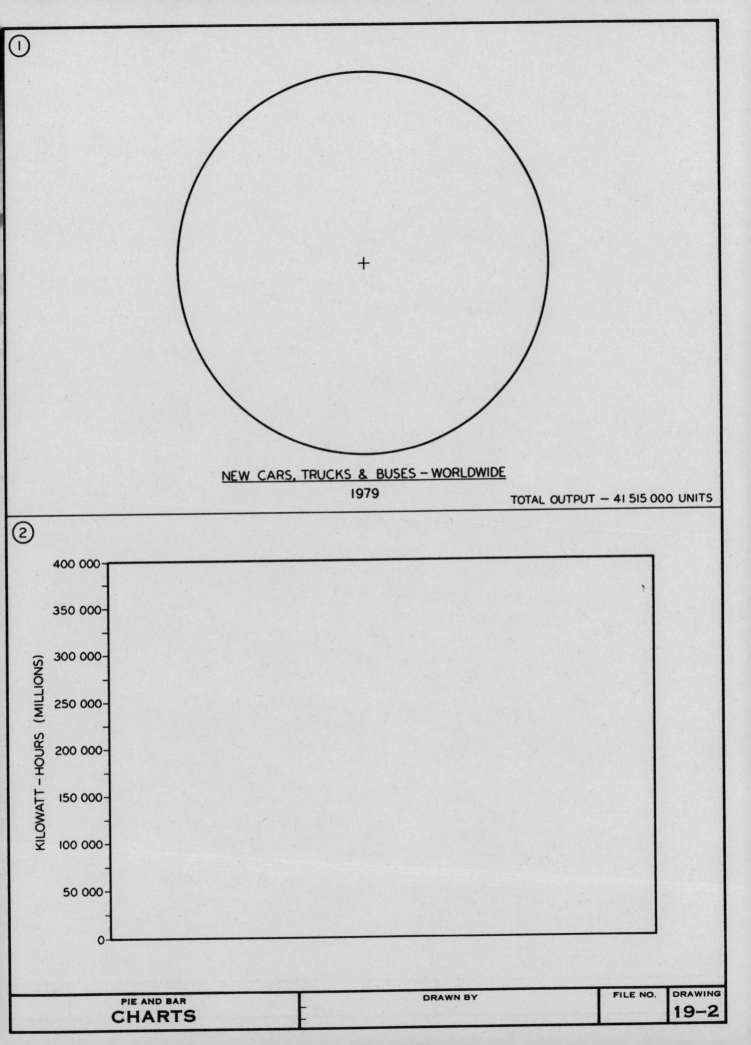

<u>NEW CARS, TRUCKS & BUSES — WORLDWIDE</u>

1979

TOTAL OUTPUT — 41 515 000 UNITS

②

KILOWATT – HOURS (MILLIONS)

400 000

350 000

300 000

250 000

200 000

150 000

100 000

50 000

0

| PIE AND BAR CHARTS | DRAWN BY | FILE NO. | DRAWING 19–2 |

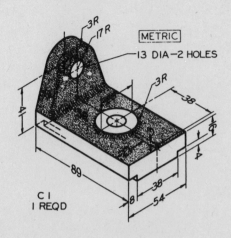

1 CUT-OFF HOLDER

METRIC

3R
17R
13 DIA – 2 HOLES
3R
38
89
38
54
8
C 1
1 REQD

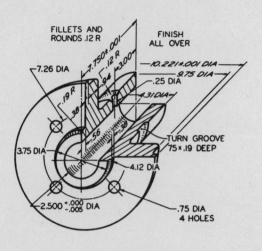

2 BEARING

FILLETS AND ROUNDS .12 R
FINISH ALL OVER
7.26 DIA
.750 ± .001
.12 R
3.00
.94
10.221 ± .001 DIA
9.75 DIA
.25 DIA
4.31 DIA
.19 R
38
.56
38
TURN GROOVE 75 × .19 DEEP
3.75 DIA
4.12 DIA
2.500 +.000 -.005 DIA
.75 DIA 4 HOLES

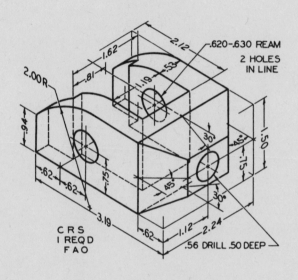

3 CAM

2.00 R
1.62
.81
2.12
.620–.630 REAM 2 HOLES IN LINE
1.19
.53
.94
30°
1.50
.75
.45
.62
.62
.75
45°
30°
3.19
.62
1.12
2.24
.56 DRILL .50 DEEP
C R S
1 REQ D
F A O

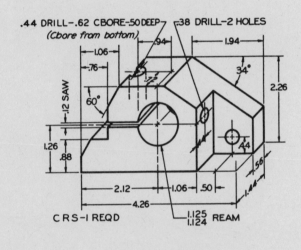

4 CLUTCH LEVER

.44 DRILL–.62 CBORE–.50 DEEP (Cbore from bottom)
.38 DRILL–2 HOLES
1.06
.94
1.94
.76
34°
.12 SAW
60°
2.26
1.26
.88
.44
2.12
1.06
.50
.56
.44
4.26
1.125 1.124 REAM
C R S–1 REQD

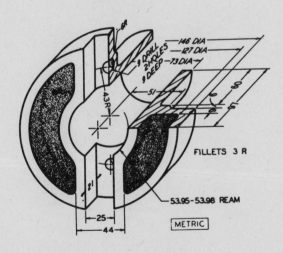

5 STOCK GUIDE

6R
146 DIA
127 DIA
9 DRILL 2 HOLES 9 DEEP
73 DIA
43 R
51
FILLETS 3 R
21
53.95–53.98 REAM
25
44
METRIC

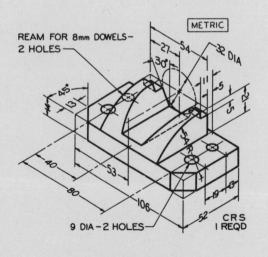

6 LOCATING FINGER

METRIC
REAM FOR 8mm DOWELS– 2 HOLES
27
54
32 DIA
30°
11
5
45°
13
5
44
40
53
80
106
9 DIA – 2 HOLES
19
13
52
C R S
1 REQD

DETAIL DRAWINGS

DRAW OR SKETCH THE NECESSARY VIEWS OF THE OBJECT ASSIGNED. DIMENSION COMPLETELY

DRAWING
20–1

DRAWN BY

FILE NO.

DRAWING

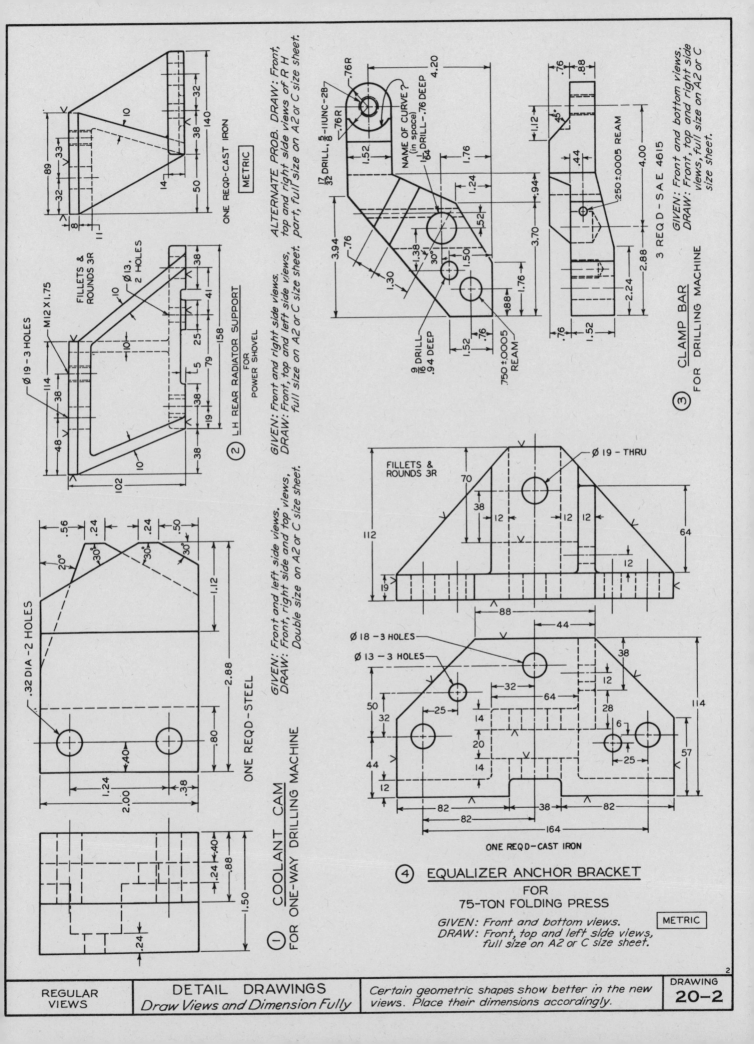

ONE REQD-CAST IRON

METRIC

ALTERNATE PROB. DRAW: Front,
top and right side views of R H
part, full size on A2 or C size sheet.

FILLETS &
ROUNDS 3R

Ø13.
2 HOLES

M12×1.75

Ø19 - 3 HOLES

② LH REAR RADIATOR SUPPORT
FOR
POWER SHOVEL

GIVEN: Front and right side views.
DRAW: Front, top and left side views,
full size on A2 or C size sheet.

NAME OF CURVE?
(in space)

17/32 DRILL, 5/8-11UNC-2B

.76R

.76R

1/64 DRILL -.76 DEEP

9/16 DRILL
.94 DEEP

.750±.0005
REAM

③ CLAMP BAR
FOR DRILLING MACHINE

3 REQD-SAE 4615

.250±.0005 REAM

GIVEN: Front and bottom views.
DRAW: Front, top and right side
views, full size on A2 or C
size sheet.

.32 DIA-2 HOLES

ONE REQD-STEEL

GIVEN: Front and left side views.
DRAW: Front, right side and top views,
Double size on A2 or C size sheet.

① COOLANT CAM
FOR ONE-WAY DRILLING MACHINE

FILLETS &
ROUNDS 3R

Ø 19 - THRU

Ø 18 - 3 HOLES

Ø 13 - 3 HOLES

ONE REQD-CAST IRON

④ EQUALIZER ANCHOR BRACKET
FOR
75-TON FOLDING PRESS

METRIC

GIVEN: Front and bottom views.
DRAW: Front, top and left side views,
full size on A2 or C size sheet.

| REGULAR VIEWS | DETAIL DRAWINGS
Draw Views and Dimension Fully | Certain geometric shapes show better in the new views. Place their dimensions accordingly. | DRAWING
20-2 |

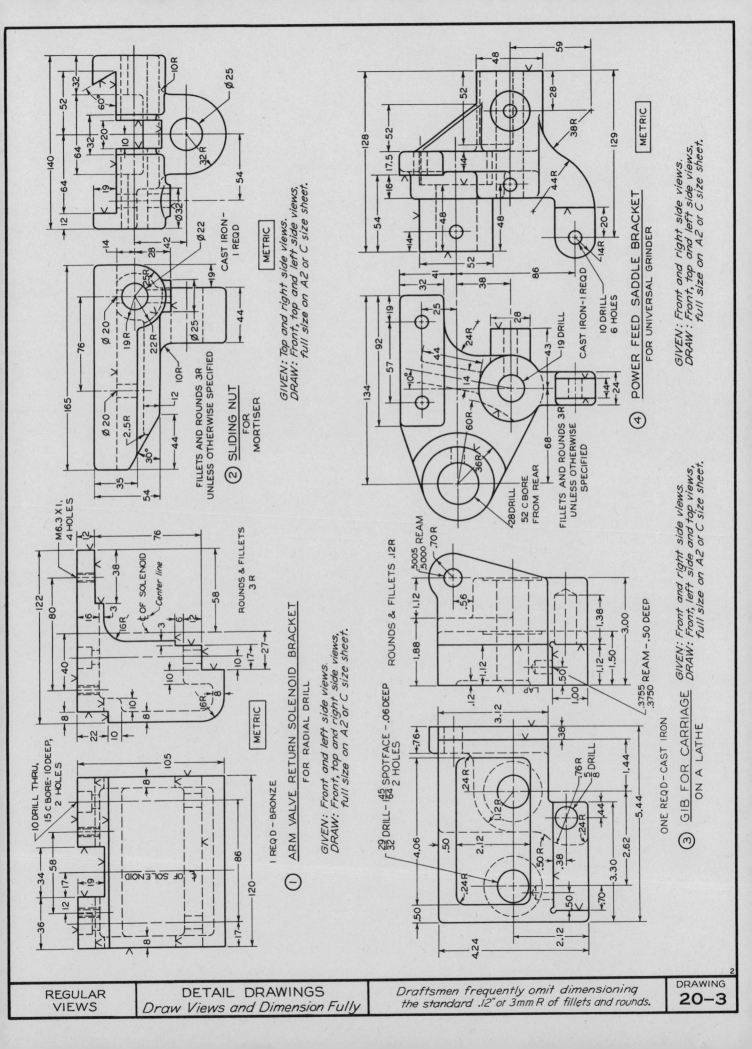

METRIC

② SLIDING NUT
FOR
MORTISER

FILLETS AND ROUNDS 3R
UNLESS OTHERWISE SPECIFIED

CAST IRON—
1 REQD

GIVEN: Top and right side views.
DRAW: Front, top and left side views,
full size on A2 or C size sheet.

METRIC

① ARM VALVE RETURN SOLENOID BRACKET
FOR RADIAL DRILL

1 REQD - BRONZE

GIVEN: Front and left side views.
DRAW: Front, top and right side views,
full size on A2 or C size sheet.

METRIC

④ POWER FEED SADDLE BRACKET
FOR UNIVERSAL GRINDER

CAST IRON-1 REQD

GIVEN: Front and right side views.
DRAW: Front, top and left side views,
full size on A2 or C size sheet.

FILLETS AND ROUNDS 3R
UNLESS OTHERWISE
SPECIFIED

---

③ GIB FOR CARRIAGE
ON A LATHE

ONE REQD-CAST IRON

GIVEN: Front and right side views.
DRAW: Front, left side and top views,
full size on A2 or C size sheet.

ROUNDS & FILLETS .12R

| REGULAR VIEWS | DETAIL DRAWINGS
Draw Views and Dimension Fully | *Draftsmen frequently omit dimensioning
the standard .12" or 3mm R of fillets and rounds.* |

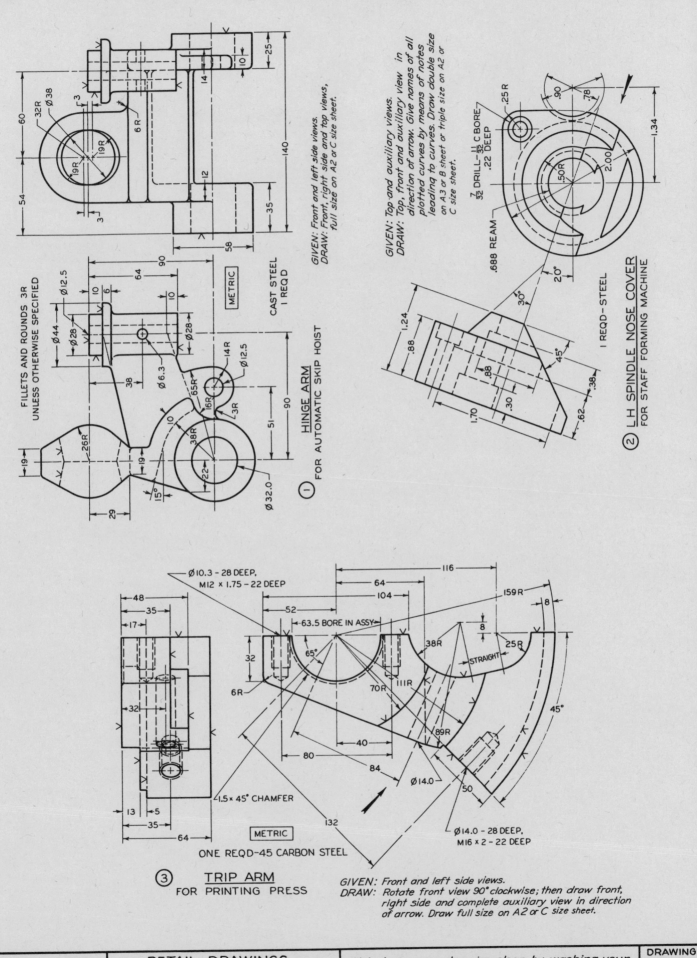

GIVEN: Front and left side views.
DRAW: Front, right side and top views, full size on A2 or C size sheet.

FILLETS AND ROUNDS 3R UNLESS OTHERWISE SPECIFIED

METRIC

CAST STEEL
1 REQ'D

① HINGE ARM
FOR AUTOMATIC SKIP HOIST

GIVEN: Top and auxiliary views.
DRAW: Top, front and auxiliary view in direction of arrow. Give names of all plotted curves by means of notes leading to curves. Draw double size on A3 or B sheet or triple size on A2 or C size sheet.

7/32 DRILL-11/32 C BORE
.22 DEEP

.688 REAM

② L H SPINDLE NOSE COVER
FOR STAFF FORMING MACHINE

1 REQ'D-STEEL

Ø10.3 - 28 DEEP,
M12 × 1.75 - 22 DEEP

63.5 BORE IN ASSY

STRAIGHT

6R

1.5 × 45° CHAMFER

Ø14.0

Ø14.0 - 28 DEEP,
M16 × 2 - 22 DEEP

METRIC

ONE REQ'D-45 CARBON STEEL

③ TRIP ARM
FOR PRINTING PRESS

GIVEN: Front and left side views.
DRAW: Rotate front view 90° clockwise; then draw front, right side and complete auxiliary view in direction of arrow. Draw full size on A2 or C size sheet.

| PLOTTED CURVES AND AUX VIEWS | DETAIL DRAWINGS Draw Views and Dimension Fully | Help keep your drawing clean by washing your hands and removing your coat or sweater. | DRAWING 20-4 |

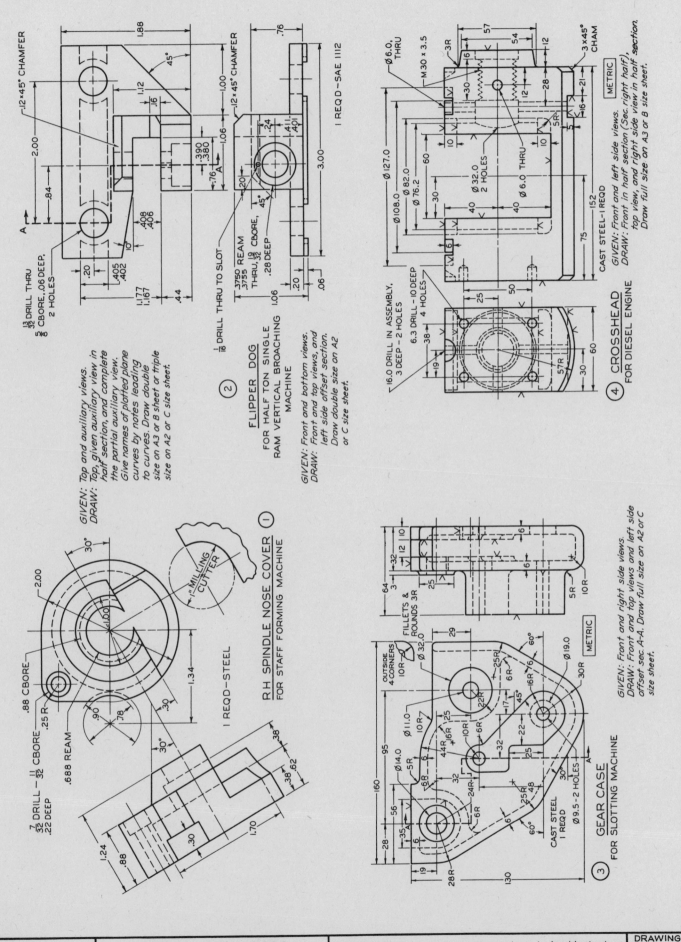

① R.H SPINDLE NOSE COVER
FOR STAFF FORMING MACHINE

1 REQD—STEEL

2.00
.88 CBORE
.25 R
.688 REAM
.90
.78
.30
30°
30°
1.00
1.34
¼" MILLING CUTTER
7/32 DRILL — 11/32 CBORE
.22 DEEP
1.24
.88
.30
1.70
.38
.62
.38

GIVEN: Top and auxiliary views.
DRAW: Top, given auxiliary view in half section, and complete the partial auxiliary view. Give names of plotted plane curves by notes leading to curves. Draw double size on A3 or B sheet or triple size on A2 or C size sheet.

② FLIPPER DOG
FOR HALF TON SINGLE RAM VERTICAL BROACHING MACHINE

.12×45° CHAMFER
1.88
45°
1.12
.16
1.00
2.00
.84
A
.20
.408/.406
.405/.402
.10
1.77/1.67
.44
13/32 DRILL THRU
5/8 CBORE,.06 DEEP,
2 HOLES
.12×45° CHAMFER
.76
1.06
.390/.380
.24
.20
45°
.76
A
4.11
4.01
3.00
.3750/.3755 REAM THRU, 19/32 CBORE, .28 DEEP
1.06
.20
.06
1/16 DRILL THRU TO SLOT
1 REQD—SAE 1112

GIVEN: Front and bottom views.
DRAW: Front and top views, and left side offset section. Draw double size on A2 or C size sheet.

③ GEAR CASE
FOR SLOTTING MACHINE

METRIC
64
12
10
3
32
25
6
6
5R
10R
FILLETS & ROUNDS 3R
OUTSIDE 4 CORNERS 10R
Ø 32.0
29
60°
25R
6R
16R 6
45°
22
32
25
30R
160
95
Ø 11.0
10R 7
25
44R 16R
10R
6R
Ø 14.0
5R
6
32
24R
6R
6R
60°
25R
48
Ø 19.0
17
22R
28
56
35
6
19
28R
130
Ø 9.5—2 HOLES
CAST STEEL 1 REQD
A

GIVEN: Front and right side views.
DRAW: Front and top views and left side offset sec. A-A. Draw full size on A2 or C size sheet.

④ CROSSHEAD
FOR DIESEL ENGINE

57
54
12
3R
6
M30 × 3.5
3×45° CHAM
30
12
28
21
16
5
10
5R
10
Ø 6.0 THRU
Ø 127.0
Ø 108.0
Ø 82.0
Ø 76.2
Ø 32.0 2 HOLES
Ø 6.0 THRU
60
30
40
40
75
152
CAST STEEL—1 REQD
METRIC
16
16.0 DRILL IN ASSEMBLY, 3 DEEP—2 HOLES
6.3 DRILL—10 DEEP 4 HOLES
25
50
38
19
57R
30
60

GIVEN: Front and left side views.
DRAW: Front in half section (Sec. right half), top view, and right side view in half section. Draw full size on A3 or B size sheet.

| AUXILIARY VIEWS AND SECTIONS | DETAIL DRAWINGS *Draw Views and Dimension Fully* | *Sectional views may be dimensioned. Avoid placing dimensions within the hatched areas, if possible.* | DRAWING 20-5 |

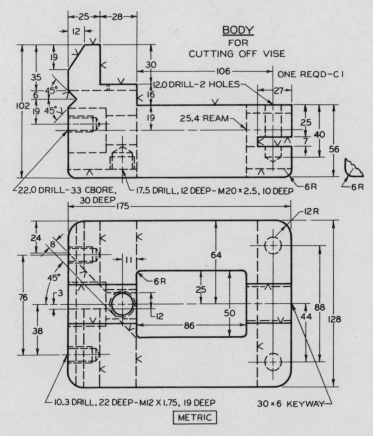

BODY
FOR
CUTTING OFF VISE

ONE REQD—C1

12.0 DRILL—2 HOLES

25.4 REAM

22.0 DRILL—33 CBORE, 30 DEEP

17.5 DRILL, 12 DEEP—M20 × 2.5, 10 DEEP

6R

6R

10.3 DRILL, 22 DEEP—M12 × 1.75, 19 DEEP

30 × 6 KEYWAY

METRIC

① GIVEN: Front and bottom views.
DRAW: Front full section, top and right side views.
Draw full size on A2 or C size sheet.

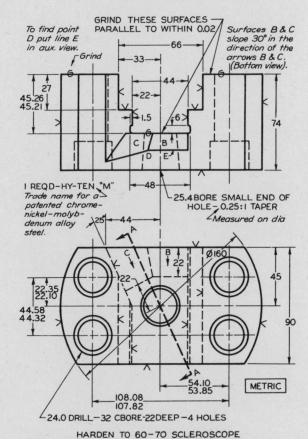

GRIND THESE SURFACES PARALLEL TO WITHIN 0.02

Surfaces B & C slope 30° in the direction of the arrows B & C. (Bottom view).

To find point D put line E in aux. view.

Grind

1 REQD—HY-TEN "M"
Trade name for a patented chrome-nickel-molyb-denum alloy steel.

25.4 BORE SMALL END OF HOLE—0.25:1 TAPER
Measured on dia

Ø160

METRIC

24.0 DRILL—32 CBORE-22DEEP—4 HOLES

HARDEN TO 60-70 SCLEROSCOPE

The scleroscope is a hardness tester based on the rebound of a diamond-tipped hammer striking the metal being tested. The "60-70" is a reading on a special scale.

STRIPPER BLOCK
FOR FIXTURE ON 25-TON 2-COLUMN PRESS

② GIVEN: Front and bottom views.
DRAW: Front, top, auxiliary section A-A, and right side view in full section.
Draw full size on A2 or C size sheet.

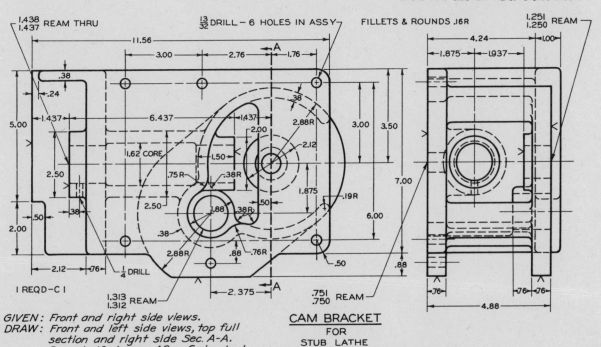

1.438 / 1.437 REAM THRU

13/32 DRILL — 6 HOLES IN ASSY

FILLETS & ROUNDS .16R

1.251 / 1.250 REAM

1.62 CORE

1.437

.75 R

2.88R

.38R

1/4 DRILL

1.313 / 1.312 REAM

2.88R

.76R

.751 / .750 REAM

1 REQD—C1

③ GIVEN: Front and right side views.
DRAW: Front and left side views, top full section and right side Sec. A-A.
Draw half size on A2 or C size sheet.

CAM BRACKET
FOR
STUB LATHE

SECTIONS

DETAIL DRAWINGS
Draw Views and Dimension Fully

Certain dimensions belong on the new view where certain geometric shapes will show best.

DRAWING
20-6

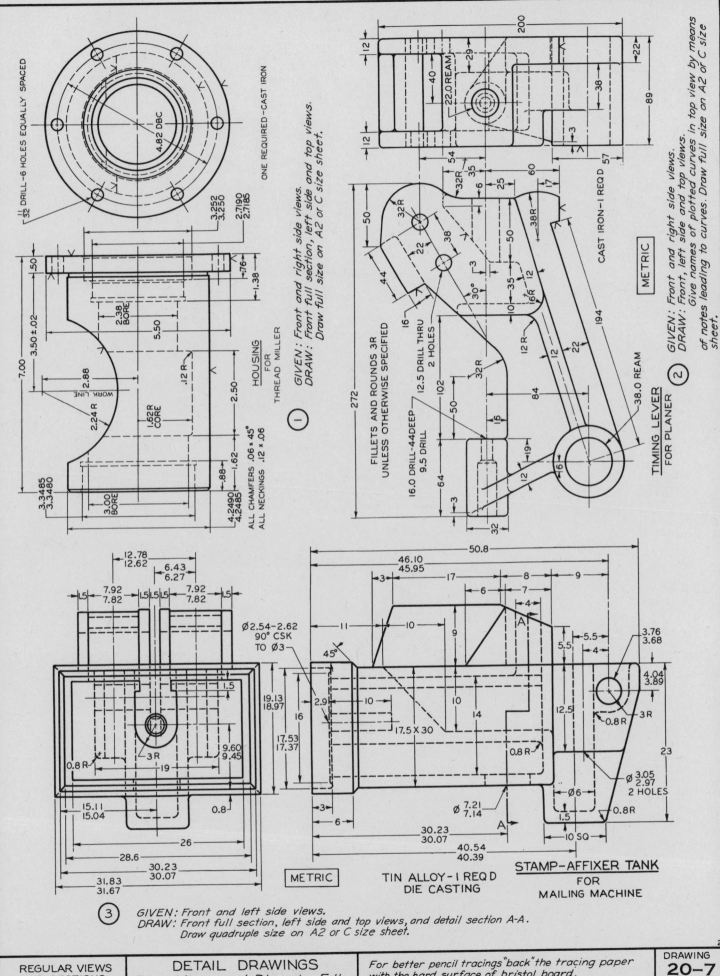

① GIVEN: Front and right side views.
DRAW: Front full section, left side and top views.
Draw full size on A2 or C size sheet.

HOUSING
FOR
THREAD MILLER

ONE REQUIRED—CAST IRON

11/32 DRILL-6 HOLES EQUALLY SPACED

4.82 DBC

3.252
3.250

2.7190
2.7185

.50
.76
1.38

2.38 BORE
5.50
.12 R

7.00
3.50 ± .02
2.88
WORK LINE
2.24 R
1.62R CORE
2.50
.88
1.62
4.2490
4.2485

3.3485
3.3480
3.00 BORE

ALL CHAMFERS .06 × 45°
ALL NECKINGS .12 × .06

② GIVEN: Front and right side views.
DRAW: Front, left side and top views.
Give names of plotted curves in top view by means
of notes leading to curves. Draw full size on A2 or C size
sheet.

TIMING LEVER
FOR PLANER

CAST IRON—I REQ D

METRIC

FILLETS AND ROUNDS 3R
UNLESS OTHERWISE SPECIFIED

16.0 DRILL-44DEEP
9.5 DRILL

12.5 DRILL THRU
2 HOLES

200
12
40
22.0 REAM
29
22
89
38
12
3
54
57
35
6
25
60
17
32R
50
22
38
3
30°
35
12
6R
32R
194
12
22
38.0 REAM
102
32R
16
50
84
12
19
6
272
44
16
50
64
3
32

③ GIVEN: Front and left side views.
DRAW: Front full section, left side and top views, and detail section A-A.
Draw quadruple size on A2 or C size sheet.

METRIC

TIN ALLOY-I REQ D
DIE CASTING

STAMP-AFFIXER TANK
FOR
MAILING MACHINE

12.78
12.62
6.43
6.27
7.92
7.82
1.5
1.5
1.5
7.92
7.82
1.5

Ø2.54-2.62
90° CSK
TO Ø3

1.5
19.13
18.97
16
17.53
17.37
9.60
9.45
3R
0.8 R
19
15.11
15.04
0.8
26
28.6
30.23
30.07
31.83
31.67

50.8
46.10
45.95
3
17
8
9
6
7
4
11
10
9
A
5.5
5.5
4
45°
2.9
10
10
14
17.5 × 30
12.5
0.8 R
3R
0.8 R
4.04
3.89
3.76
3.68
23
Ø 7.21
7.14
A
Ø6
Ø 3.05
2.97
2 HOLES
3
6
30.23
30.07
40.54
40.39
1.5
10 SQ
0.8R

REGULAR VIEWS
AND SECTIONS

DETAIL DRAWINGS
Draw Views and Dimension Fully

For better pencil tracings "back" the tracing paper
with the hard surface of bristol board.

DRAWING
20-7

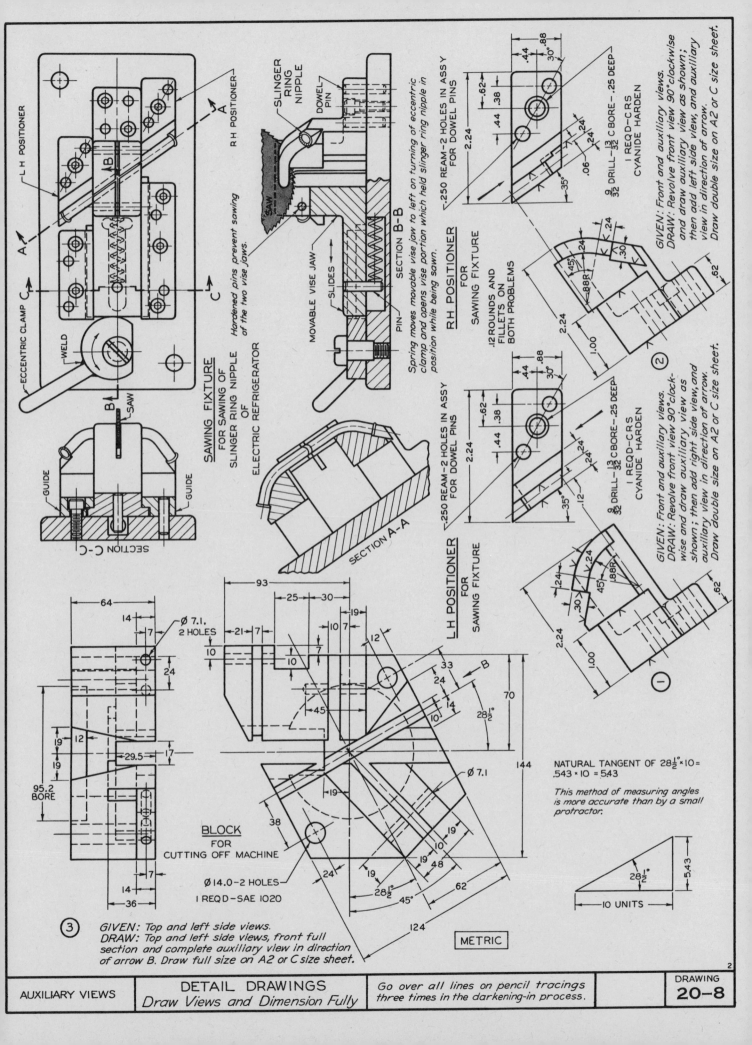

LH POSITIONER

RH POSITIONER

ECCENTRIC CLAMP C

WELD

A

B

SAW

GUIDE GUIDE

SECTION C-C

SAWING FIXTURE
FOR SAWING OF
SLINGER RING NIPPLE
OF
ELECTRIC REFRIGERATOR

Hardened pins prevent sawing
of the two vise jaws.

SLINGER
RING
NIPPLE

DOWEL
PIN

SAW

MOVABLE VISE JAW

SLIDES

PIN

SECTION B-B

Spring moves movable vise jaw to left on turning of eccentric
clamp and opens vise portion which held slinger ring nipple in
position while being sawn.

SECTION A-A

∼.250 REAM-2 HOLES IN ASS'Y
FOR DOWEL PINS

.88
.44
30°
.62
.38
.44
2.24
.24
.24
.06
35°

9/32 DRILL - 13/32 C BORE - .25 DEEP
I REQ'D - CRS
CYANIDE HARDEN

GIVEN: Front and auxiliary views.
DRAW: Revolve front view 90° clockwise
and draw auxiliary view as shown;
then add left side view, and auxiliary
view in direction of arrow.
Draw double size on A2 or C size sheet.

RH POSITIONER
FOR
SAWING FIXTURE

.24
.24
.30
45°
.88R
2.24
1.00
.62

②

∼.250 REAM-2 HOLES IN ASS'Y
FOR DOWEL PINS

.88
.44
30°
.62
.38
.44
2.24
.24
.24
.12
35°

9/32 DRILL - 13/32 C BORE - .25 DEEP
I REQ'D - CRS
CYANIDE HARDEN

GIVEN: Front and auxiliary views.
DRAW: Revolve front view 90° clockwise and draw auxiliary view as
shown; then add right side view, and
auxiliary view in direction of arrow.
Draw double size on A2 or C size sheet.

LH POSITIONER
FOR
SAWING FIXTURE

.24
.24
.30
45°
.88R
2.24
1.00
.62

①

64
14
7
Ø 7.1,
2 HOLES
24
19
12
29.5
17
19
95.2
BORE
7
14
36

BLOCK
FOR
CUTTING OFF MACHINE

Ø 14.0 - 2 HOLES
I REQ'D - SAE 1020

93
25
30
21 7
10 7
19
10
10
7
12
45
33
24
14
10
28½°
70
144

Ø 7.1

38
19
19
10
19
48
24
19
62
28½°
45°
124

NATURAL TANGENT OF 28½° × 10 =
.543 × 10 = 5.43

This method of measuring angles
is more accurate than by a small
protractor.

28½°
5.43
10 UNITS

METRIC

③ GIVEN: Top and left side views.
DRAW: Top and left side views, front full
section and complete auxiliary view in direction
of arrow B. Draw full size on A2 or C size sheet.

.12 ROUNDS AND
FILLETS ON
BOTH PROBLEMS

| AUXILIARY VIEWS | DETAIL DRAWINGS | Go over all lines on pencil tracings | DRAWING |
| | Draw Views and Dimension Fully | three times in the darkening-in process. | 20-8 |

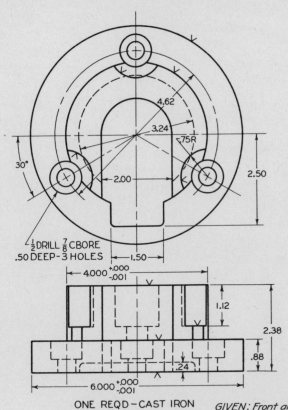

½ DRILL ⅞ CBORE
.50 DEEP—3 HOLES

4.62
3.24
.75R
2.00
2.50
30°
1.50

4.000 +.000 -.001
1.12
2.38
.88
.24
6.000 +.000 -.001

ONE REQD—CAST IRON

(1) CENTERING BUSHING
FOR A KEY SEATER

GIVEN: Front and bottom views.
DRAW: Front and left side views, and right side full section. Draw full size on A3 or B size sheet.

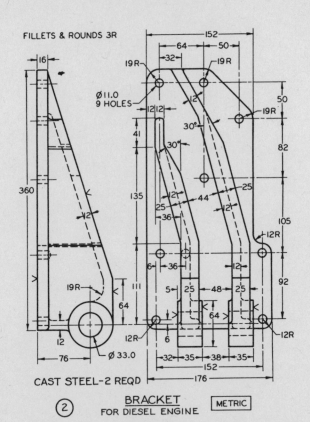

FILLETS & ROUNDS 3R

16
152
64 50
32
19R 19R
19R
Ø11.0 9 HOLES
12 12
41
30°
50
82
30°
30°
360
135
25
36
12 44 12 25
111
25 48 25
19R
64
12R
12R 12R
76
Ø 33.0
32 35 38 35
152
176
5 25 12
92

CAST STEEL—2 REQD

(2) BRACKET
FOR DIESEL ENGINE   METRIC

GIVEN: Front and left side views.
DRAW: Revolve front view 90° clockwise and let it be a top view; then add front and right side views. Draw half size on A3 or B size sheet.

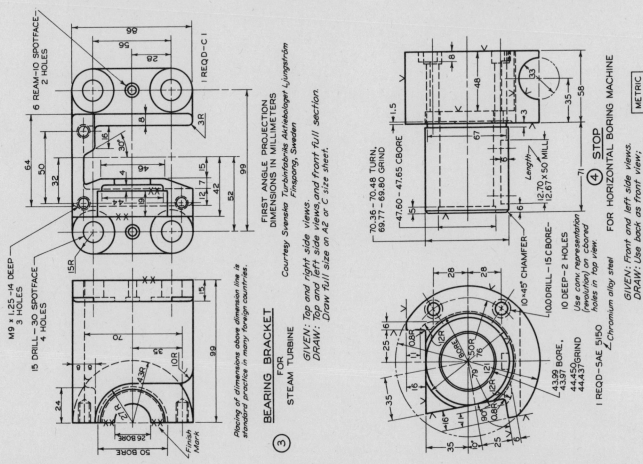

6 REAM—10 SPOTFACE 2 HOLES
86
56
28
8
3R
16
30°
64
50
46
4
32
12 7 15
42
52
99
99
44
M9 × 1.25—14 DEEP 3 HOLES
15 DRILL—30 SPOTFACE 4 HOLES
15R
15
70
35
10R
43R
2R
24
26 BORE
50 BORE
Finish Mark

1 REQD—C I

FIRST ANGLE PROJECTION
DIMENSIONS IN MILLIMETERS

Courtesy Svenska Turbinfabriks Aktiebolaget Ljungström
Finspong, Sweden

GIVEN: Top and right side views.
DRAW: Top and left side views, and front full section. Draw full size on A2 or C size sheet.

BEARING BRACKET
FOR
STEAM TURBINE

Placing of dimensions above dimension line is standard practice in many foreign countries.

(3)

8
33
48
1.5
3
35
58
67
Length
12.70 × 50 MILL
12.67 × 50 MILL
71
6
70.36 — 70.48 TURN, 69.77 — 69.80 GRIND
47.60 — 47.65 CBORE
5
10×45° CHAMFER
10.0 DRILL—15 CBORE 10 DEEP—2 HOLES
Use conv. representation (revolution) on cbored holes in top view.

28 28
25 6
12R
0.8R
BORE 50R 76
79 121 12R
43.99 BORE, 43.97 44.450 GRIND 44.437 GRIND
35
16
35
16
90°
0.8R
12R
25 6

1 REQD—SAE 5150
Chromium alloy steel

(4) STOP
FOR HORIZONTAL BORING MACHINE   METRIC

GIVEN: Front and left side views.
DRAW: Use back as front view; then draw left side and top views. Draw full size on A2 or C size sheet.

REGULAR VIEWS AND SECTIONS

DETAIL DRAWINGS
Draw Views and Dimension Fully

Develop the several views simultaneously rather than attempt to draw each view separately.

DRAWING
20—9

2

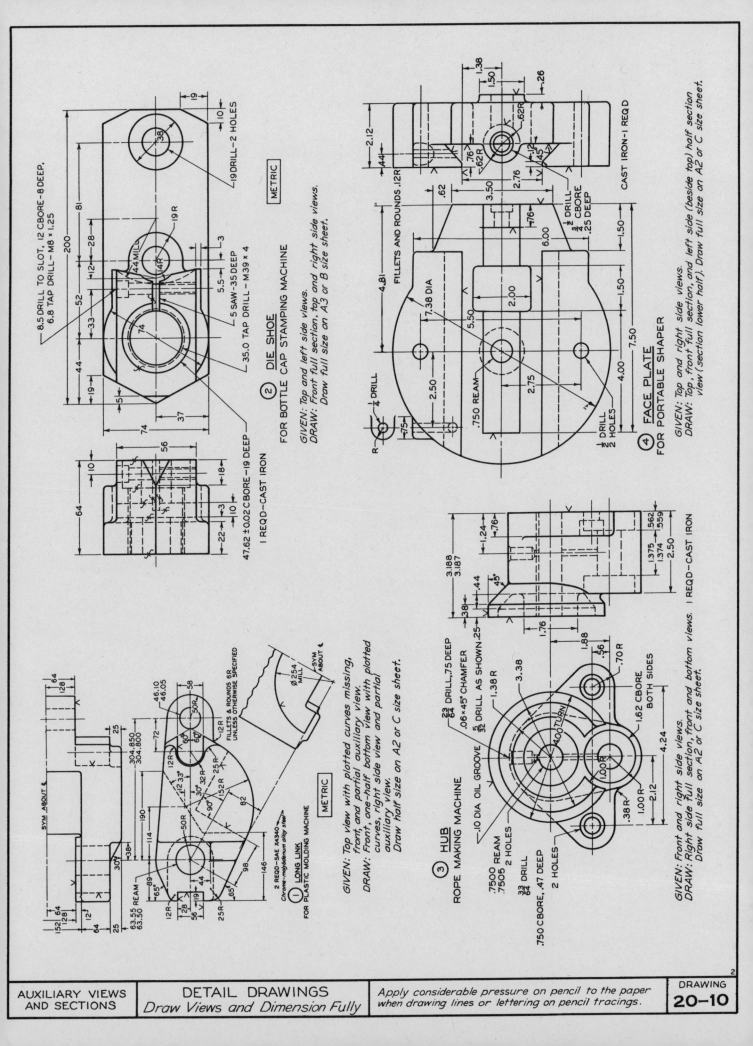

① LONG LINK
FOR PLASTIC MOLDING MACHINE

2 REQD–SAE X4340
Chrome–molybdenum alloy steel

METRIC

GIVEN: Top view with plotted curves missing,
front, and partial auxiliary view.
DRAW: Front, one-half bottom view with plotted
curves, right side view and partial
auxiliary view.
Draw half size on A2 or C size sheet.

② DIE SHOE
FOR BOTTLE CAP STAMPING MACHINE

1 REQD–CAST IRON

METRIC

GIVEN: Top and left side views.
DRAW: Front full section, top and right side views.
Draw full size on A3 or B size sheet.

③ HUB
ROPE MAKING MACHINE

1 REQD–CAST IRON

GIVEN: Front and right side views.
DRAW: Right side full section, front and bottom views.
Draw full size on A2 or C size sheet.

④ FACE PLATE
FOR PORTABLE SHAPER

CAST IRON–1 REQD

GIVEN: Top and right side views.
DRAW: Top, front full section, and left side (beside top) half section
view (section lower half). Draw full size on A2 or C size sheet.

FILLETS AND ROUNDS .12R

| AUXILIARY VIEWS AND SECTIONS | DETAIL DRAWINGS *Draw Views and Dimension Fully* | *Apply considerable pressure on pencil to the paper when drawing lines or lettering on pencil tracings.* | DRAWING **20-10** |

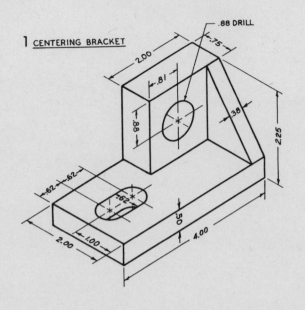

1 CENTERING BRACKET

.88 DRILL
2.00
.75
.81
.88
.38
2.25
.62 .62
.62
.50
2.00
1.00
4.00

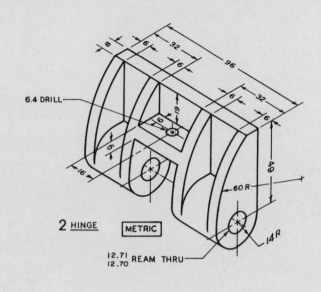

2 HINGE   METRIC

8
6
32
96
6
6
32
6.4 DRILL
6
19
6
49
16
60 R
12.71
12.70 REAM THRU
14 R

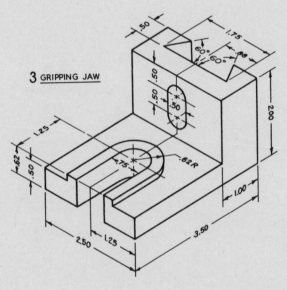

3 GRIPPING JAW

.50
1.75
60° 60°
.38
.50
.50
.50
.50
2.00
1.25
.62
.50
.75
.62 R
2.50
1.25
1.00
3.50

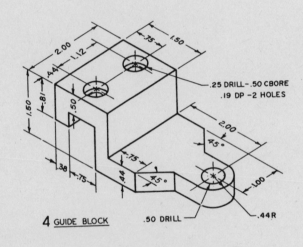

4 GUIDE BLOCK

2.00
.75
1.50
.44
1.12
.81
1.50
.50
.25 DRILL - .50 CBORE
.19 DP - 2 HOLES
2.00
4.5°
.75
.44
.38
.75
4.5°
.50 DRILL
1.00
.44 R

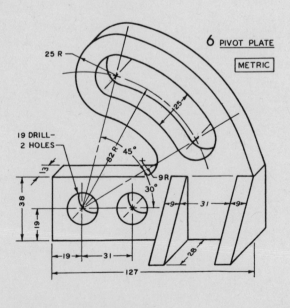

5 HINGE BASE   METRIC

27
35
13
14
60°
14
60
6
9 R
3
25
24
16
9 R
24
13
44
110
60
13.5 DRILL -
2 HOLES IN LINE

6 PIVOT PLATE   METRIC

25 R
25
19 DRILL -
2 HOLES
13
82 R
45°
9 R
30°
38
9
31
9
19
19
31
28
127

DETAIL DRAWINGS

DRAW OR SKETCH THE NECESSARY VIEWS OF THE OBJECT
ASSIGNED. DIMENSION COMPLETELY

DRAWING
20-11

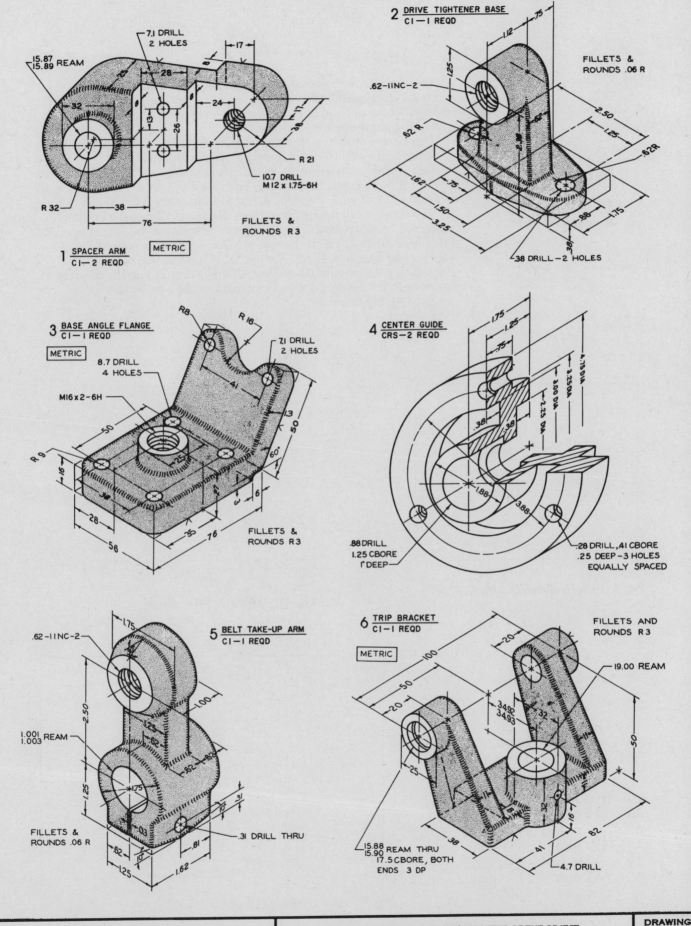

1 SPACER ARM
C1— 2 REQD
METRIC

15.87 REAM
15.89 REAM
32
R 32
38
76
7.1 DRILL
2 HOLES
28
13
26
24
17
8
38
17
R 21
10.7 DRILL
M12 x 1.75-6H
FILLETS &
ROUNDS R 3

2 DRIVE TIGHTENER BASE
C1—1 REQD

1.12
.75
1.25
.62-11NC-2
.62
82 R
2.50
1.25
82 R
1.62
.75
1.50
3.25
.88
1.75
.38
.38 DRILL—2 HOLES
FILLETS &
ROUNDS .06 R

3 BASE ANGLE FLANGE
C1—1 REQD
METRIC

R 8
R 16
7.1 DRILL
2 HOLES
8.7 DRILL
4 HOLES
M16 x 2-6H
41
13
50
50
R 9
60°
16
25
6
3
28
35
76
56
FILLETS &
ROUNDS R 3

4 CENTER GUIDE
CRS-2 REQD

1.75
1.25
.75
4.75 DIA
3.00 DIA
2.25 DIA
2.25 DIA
.38
.38
1.88
3.88
.88 DRILL
1.25 CBORE
1" DEEP
.28 DRILL, .41 CBORE
.25 DEEP—3 HOLES
EQUALLY SPACED

5 BELT TAKE-UP ARM
C1—1 REQD

1.75
.62-11NC-2
2.50
1.00
1.001 REAM
1.003 REAM
1.25
.62
.82
.83
1.25
.175
.31
.10
.03
FILLETS &
ROUNDS .06 R
.62
.10
1.25
.81
1.62
.31 DRILL THRU

6 TRIP BRACKET
C1—1 REQD
METRIC

FILLETS AND
ROUNDS R 3
20
100
50
20
19.00 REAM
34.92
34.93
32
50
25
32
16
15.88 REAM THRU
15.90
17.5 CBORE, BOTH
ENDS 3 DP
38
41
82
4.7 DRILL

DETAIL DRAWINGS

DRAW OR SKETCH THE NECESSARY VIEWS OF THE OBJECT
ASSIGNED. DIMENSION COMPLETELY

DRAWING
20-12

Sheet Layouts

A convenient code to identify American National Standard sheet sizes and forms suggested by the authors for title, parts or material list, and revision blocks, for use of instructors in making assignments, is shown here. All dimensions are in inches.

Three **sizes** of sheets are illustrated: **Size A**, Fig. I, **Size B**, Fig. V, and **Size C**, Fig. VI. Metric size sheets are not shown.

Eight **forms** of lettering arrangements are suggested, known as **Forms 1, 2, 3, 4, 5, 6, 7,** and **8**, as shown below and opposite. The total length of **Forms 1, 2, 3,** and **4** may be adjusted to fit **Sizes A4, A3,** and **A2**.

The term **layout** designates a sheet of certain size plus a certain arrangement of lettering. Thus **Layout A–1** is a combination of **Size A**, Fig. I, and **Form 1**, Fig. II. **Layout C–678** is a combination of **Size C**, Fig. VI, and **Forms 6, 7,** and **8**, Figs. IX, X, and XI. **Layout A4–2** (adjusted) is a combination of **Size A4** and **Form 2**, Fig. III, adjusted to fit between the borders. Other combinations may be employed as assigned by the instructor.

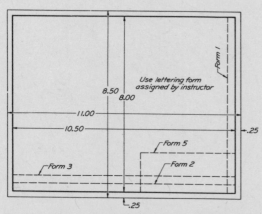

Fig. I   Size A Sheet (8.50″ × 11.00″)

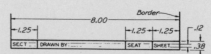

Fig. II   Form 1. Title Block

Fig. III   Form 2. Title Block

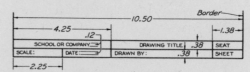

Fig. IV   Form 3. Title Block

Sheet Sizes

American National Standard

A – 8.50″ × 11.00″
B – 11.00″ × 17.00″
C – 17.00″ × 22.00″
D – 22.00″ × 34.00″
E – 34.00″ × 44.00″

International Standard

A4 – 210 mm × 297 mm
A3 – 297 mm × 420 mm
A2 – 420 mm × 594 mm
A1 – 594 mm × 841 mm
A0 – 841 mm × 1189 mm
(25.4 mm = 1.00″)

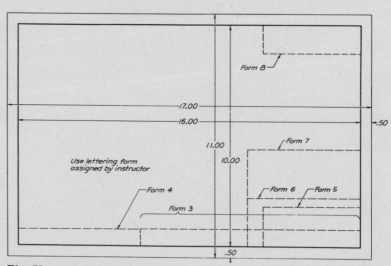

Fig. V   Size B Sheet (11.00″ × 17.00″)

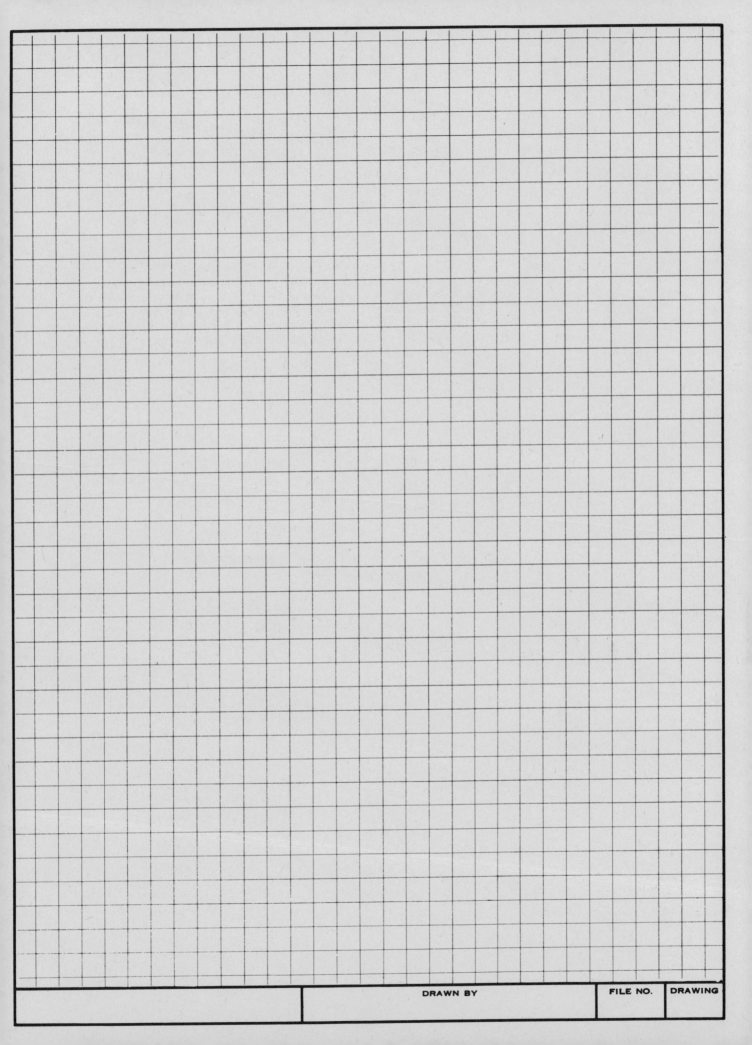

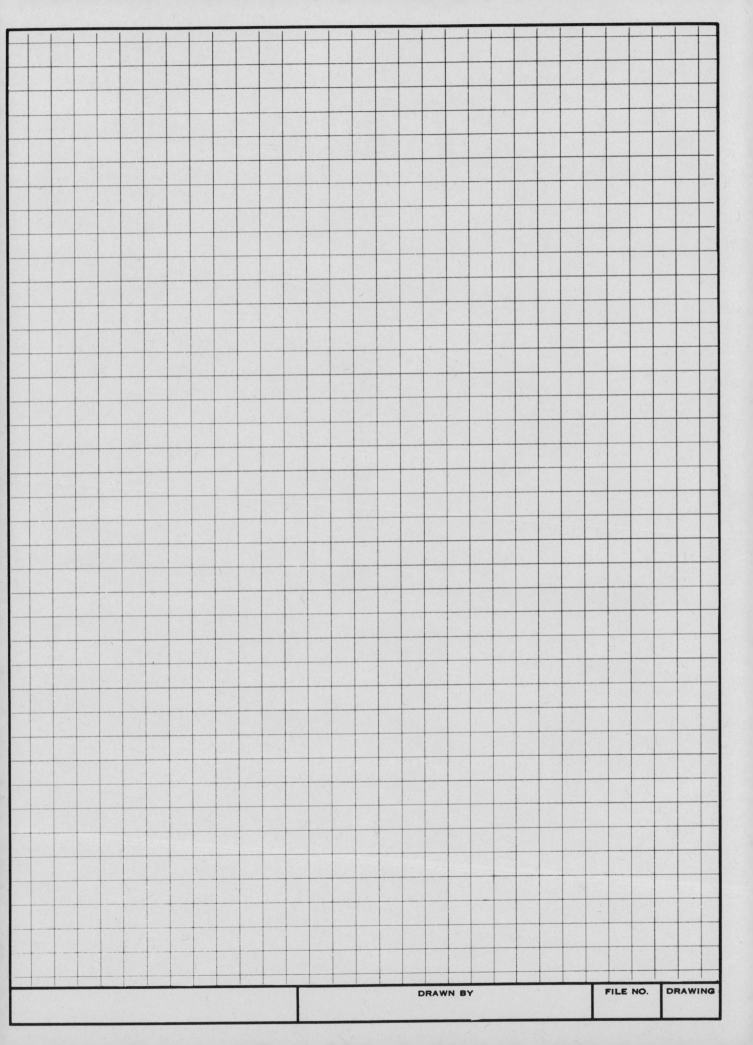

DRAWN BY   FILE NO.   DRAWING

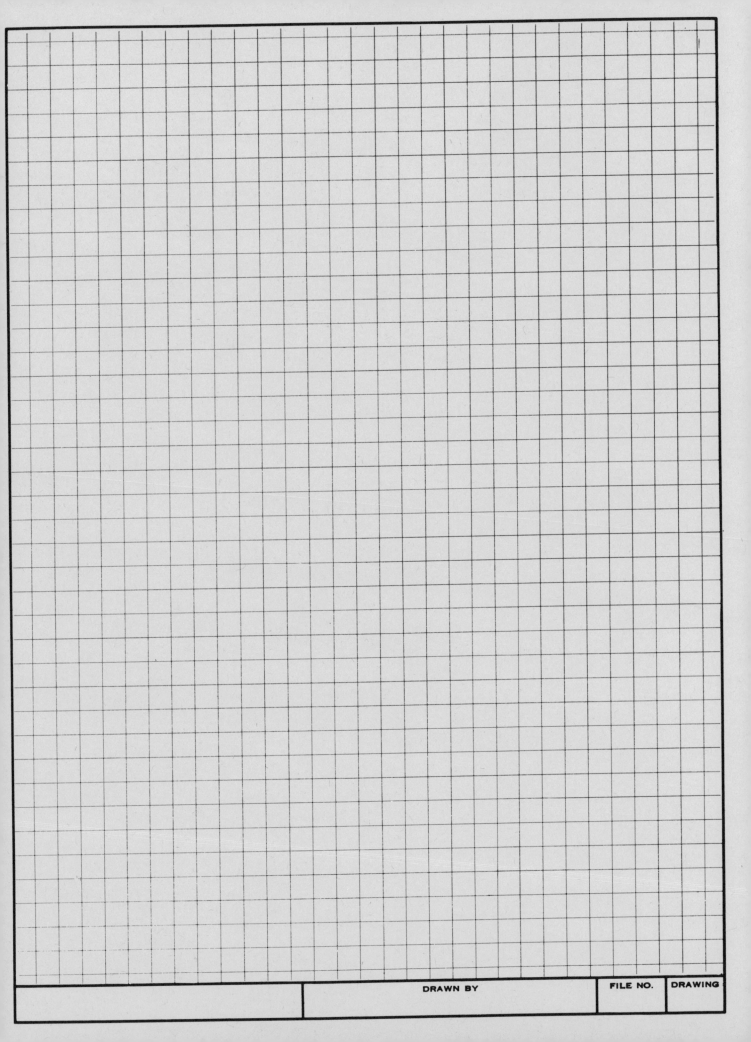

DRAWN BY

FILE NO.

DRAWING

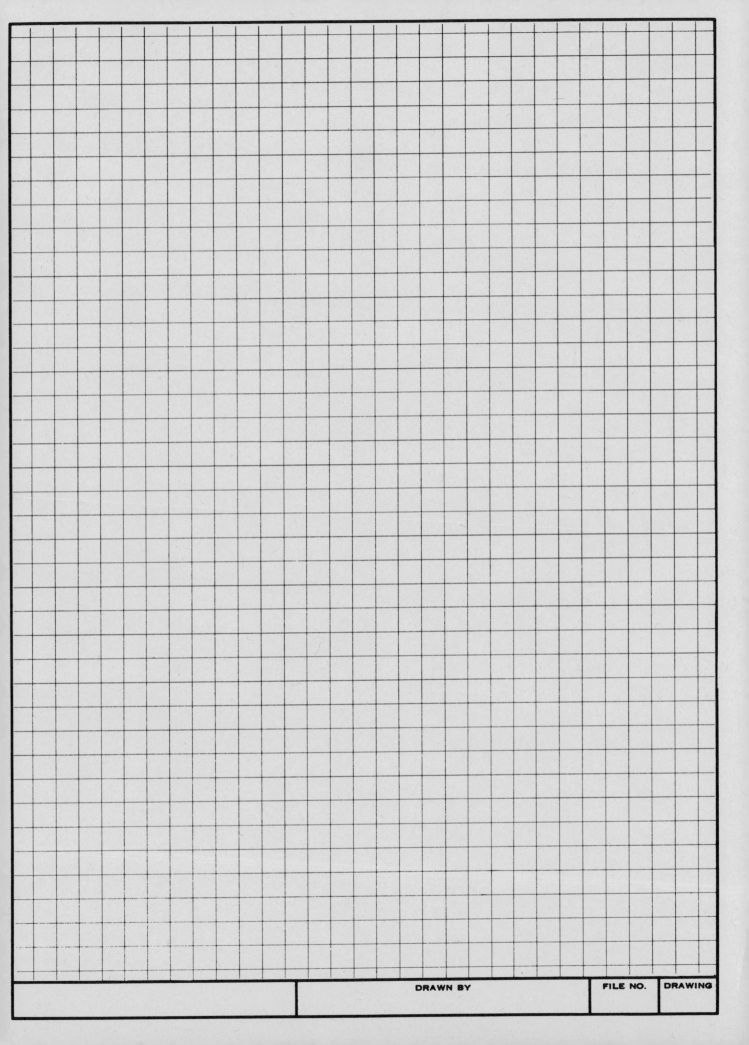

DRAWN BY

FILE NO.

DRAWING